D0852349

Lost Worlds

Lost Worlds

Bruce M. Beehler

Illustrations
by
John Anderton

Yale University Press *New Haven & London*

Adventures in the
Tropical Rainforest

Published with assistance from the foundation established in memory of
Philip Hamilton McMillan of the Class of 1894, Yale College.

Designed by Sonia Shannon.
Set in Monotype Bulmer by Duke & Company, Devon, Pennsylvania.
Printed in the United States of America.

Library of Congress Cataloging-in-Publication Data
Beehler, Bruce McP.
Lost worlds : adventures in the tropical rainforest / Bruce M. Beehler.
      p.   cm.
Includes bibliographical references and index.
ISBN 978-0-300-12228-2 (cloth : alk. paper) 1. Birds—Tropics.
2. Beehler, Bruce McP.—Travel. 3. Ornithologists—United States—Biography.
4. Naturalists—United States—Biography. I. Title.
QL695.5B44 2008
508.315′20913—dc22   2007044560

A catalogue record for this book is available from the British Library.

The paper in this book meets the guidelines for permanence and durability
of the Committee on Production Guidelines for Book Longevity
of the Council on Library Resources.

10 9 8 7 6 5 4 3 2 1

To Carol, Grace, Andrew, and Cary
who keep the home fires burning

Contents

Preface

August 28, 2000. Yongsu Camp, Papua, Indonesia. Here I was in western New Guinea. A light rain pattered on the roof of my green mountain tent. It was dark. Supper was best forgotten. I was snug in my tent, reading a tattered paperback copy of Isak Dinesen's *Out of Africa,* the pages illuminated by the narrow beam of my tiny flashlight. (Two AA cells are really not suitable to power a light for reading, but I could make do.)

I was content, for a number of reasons. I was happy because I was dry. (It's hard to stay dry in a rainforest.) I was pleased because I was at ease and comfortable. (Getting around in a rainforest often means straining and sweating *a lot.*) I was satisfied because it was night, and I could listen to the waves charging up the white coral beach. The end of a fulfilling day in the rainforest need be nothing more than this. The sound of the waves mixed with the sound of jungle katydids (*CHH CHH CHH . . . CHH CHH CHH . . . CHH CHH CHH . . .*). Even the sound of light rain was pleasant in this state of restful bliss. After I read a few more pages, sleep gathered me up, leading me to dreams of another day in the jungle.

Camping where the forest meets the beach at Yongsu had special significance for me, because at the time I was a deskbound environment officer working for the U.S. Department of State. Somehow I had sold my Foreign Service boss on letting me make a trip back to New Guinea, where I had started my rainforest studies a quarter-century earlier. Here on Yongsu beach I was far from my Washington office and its fast-accumulating emails and interagency meetings. Here at Yongsu I was playing hooky in the forest with a team of scientists from Conservation International, a Washington-based organization focused on rainforest conservation. Nothing could be more revitalizing. Thoreau was not wrong when he wrote, "In wildness is the preservation of the world." It is also salvation for the soul wearied by official briefing papers and multilateral negotiations.

I was gratified to be once again in the tropical jungle. It wasn't because of the food, certainly. The cook at our Yongsu camp seemed to have a fixation on fish-head soup and white rice. I skipped the fish heads and ate white rice enlivened by a *sambal*—a spicy sauce of chilies that looked a bit like ketchup, but was fiery hot and not at all sweet. Breakfast, lunch, and dinner were an exercise in weight loss.

It wasn't because of the local trail system, which seemed designed to kill off this middle-aged ornithologist. Every trail appeared to lead steeply up, no matter which way I turned. Slipping. Puffing. Falling. Swearing. My hands stained by the orange tropical earth.

It wasn't the superb views of Papuan forest birds. My eyeglasses were usually fogged, and the birds often flew off before I could focus my binoculars.

What, then, was my motivation? It was, in part, to peer back into a time machine, for where in my Maryland homeland could I gaze out at an ancient hardwood forest? It was to visit a place where the aboriginal inhabitants had *not* been driven off their land. At Yongsu I worked with indigenous naturalists daily, and they taught me their wildlife lore. It was to be in a place where each day I needed to test my body in ways not available in the environs of Washington, D.C., except in a health club. It was the challenge of getting to understand how these complex ecosystems were put together and how they functioned. And finally, it was the absolute necessity of determining how the best of these rainforests could be preserved for generations to come—so that our great-grandchildren could hear the rainforest whisper its secrets.

The rainforest occupies a mere 7 percent of the earth's land surface, yet is home to more than half of all the described species. Many more species lurk in the rainforest, still undescribed and waiting for future naturalists to collect, name, and study them. The remarkable combination of year-round warm temperatures, constant high humidity, and absence of large cyclonic storms produce conditions conducive to the evolution of an astoundingly rich forest environment. The great physical and taxonomic complexity of the rainforest plant life creates a diversity of habitats for forest-dwelling animal forms, both large and small. Perhaps most amazing is the degree of ecological specialization that takes place there, and the often-bizarre ecological inter-actions between species.

The rainforest environment is fragile, even though it appears imposing and impenetrable. It usually flourishes on a thin layer of topsoil that rests atop a hard base of nutrient-poor red tropical subsoil. Most of the forest's nutrients remain tied up in the wood, bark, and leaves of the trees. Once the forest has been removed, the heavy rains wash what remains of the topsoil into the rivers, leaving behind a poor substrate that, in many cases, is unable to regenerate another rainforest.

The world's remaining tropical rainforests are concentrated in the Amazon, the Congo Basin, Southeast Asia, and New Guinea. The human pressures on these forests are immense and increase every year. More than half of all the rainforests that were standing a century ago have been converted to nonforest habitats. The main threats are population growth, large-scale plantation agriculture, expanding traditional agriculture under high popula-tion densities, industrial logging, and loss of tradition among forest-dwelling cultures.

This last point is worth explaining. As forest-dwelling peoples west-ernize and modernize, they give up their forest ways and lose interest in the forests, which for millennia have provided most of the natural products those communities use for subsistence. A neglected forest is one more readily traded for cash, with the young men and women soon dispersing to the city. But does the world really need another squatter settlement in the suburbs of a big tropical city? Those who care about human welfare and about the welfare of the earth's biodiversity need to provide incentives for traditional forest peoples to stay at home in their forests and act as guardians of those forests for the generations to come. These forest stewards will be preserving their own rich cultural past and saving a resource that the world cannot do with-out—the mysterious, misunderstood, and magnificent tropical rainforest.

This book, then, is a story of the rainforest and its overlooked impor-tance to humankind's long-term well-being. It is a story I have wanted to tell for a decade, but was delayed because I was busy with rainforest research and conservation. I want to show that the rainforest is not a glorified museum of bizarre species; rather, it is a vast climate system—a source of global species diversity, and of human diversity as well.

I probably would never have traveled to the tropical world were it not for those individuals who pushed me at various stages of my life. When I was eight and struggling in school, my mother, Cary Beehler, assured me that the appreciation of birds was as important as arithmetic. She took me to the Enoch Pratt Free Library in downtown Baltimore to borrow a copy of Arthur Cleveland Bent's *Life Histories of North American Woodpeckers*. Reading these accounts set me on my first steps toward a career in ornithology. At the Adirondack Wilderness Camp, Elliot K. Verner taught me to wonder at wilderness, starting with the deep woods of northeasternmost New York State. At Williams College, Fred Rudolph taught me how to write a clear sentence and supported my bid for a Thomas J. Watson Fellowship that took me to the southwest Pacific for the first time. In graduate school at Princeton University, Henry Horn showed me how vital natural curiosity is to learning. Jared Diamond, through his scientific writings, lured me to the mountains of New Guinea and sparked my interest in ideas worth examining more closely in tropical forests. There, Thane Pratt showed me my first New Guinean forest birds on the Bulldog Road and on the slopes of Mount Missim. To all of these mentors, and others not mentioned here, my heartfelt thanks.

I also thank Carol Beehler, Tom Lovejoy, Meg Lowman, Roger Pasquier, and Fred Rudolph for reading drafts of this book and for suggesting many improvements. John Anderton drew the evocative illustrations that appear on the title pages and open the chapters. (The other figures, with two exceptions, are my own.) Jean E. Thomson Black, my editor at Yale University Press, encouraged and challenged me. Vivian Wheeler added polish to my prose. To Jean and Vivian I offer my gratitude for their patience and editorial talent.

I have capitalized the proper names of species of animals and plants so that the reader can distinguish between a Red-headed Honeyeater (a particular named species) and a red-headed honeyeater (of which there are a number of species). Thus I have capitalized Raggiana Bird of Paradise, Giant Sequoia, Black Howler Monkey, Leopard, Tiger, Leopard Cat, and many others.

I do regret that I have no chapter on the Amazon to offer, but that region has already been featured in many full-length books. I choose to focus on some lesser-known worlds—supporting actors in the great evolutionary drama played out in the earth's tropical rainforest.

Introduction

It was August 2005. Andrew, my thirteen-year-old son, and I were crammed into an overcrowded Bell 206 helicopter that was maneuvering its way through the precipitous gorge country of central Enga Province in Papua New Guinea. Rain beat against the broad, bubble-like Plexiglas windscreen, and cloud and mist were lowering onto the ridge tops. Just below us was the churning torrent of the angry Lagaip River, now the color of wet concrete. Cliffs of raw siltstone rose on our left, and regenerating forest carpeted the ravine to our right. We could see no sign of civilization or settlement—not even a garden clearing.

We were skimming over the mountainous equatorial jungles of New Guinea, an unfortunate place to be in this nasty rainstorm. Was this to be the ignoble end to our month-long field trip into the magnificent Enga uplands? I tried to put on a brave face for my son, who seemed pretty cool about our situation. My hands were clenching involuntarily. I could see no clear way ahead, because of a cloud wall to the east. Below was a beautiful but stark wilderness landscape—uninviting and forbidding. Involuntarily, I kept trying to imagine how the chopper could come down safely. It was one of a very few times airborne that I have felt a sickening knot clenching in my gut. What had I gotten my son into?

On this particular day we were on our way home from a survey of the plant and animal life in the Mount Kaijende highlands, situated in the heart of a great mountain wilderness in western Papua New Guinea. And I had come to know that in New Guinea, you take the bad with the good: a superbly productive field trip could easily be followed by a heart-pounding helicopter ride.

After about forty-five minutes of zigging and zagging through a maze of closed-in wilderness valleys, we were attempting to sneak over a pass into the open grassy valley that held our destination—the Kagamuga airstrip of Mount Hagen town. We heard the pilot mutter into the headset, "Well, we're in trouble now," and a few minutes later, "Now it's really getting dangerous."

As we flew lower and lower, and as the ground rose higher and higher toward the wall of gray cloud, we could see our window of escape closing. I thought the pilot would simply look for a clearing beside a native hamlet and set the machine down any way he could. There seemed to be no other option. But he wasn't asking for my advice; he was staring through the dismal mists. One thing we knew: we had come too far and spent too much fuel to return to base.

I had given up hope of getting to our destination. Then the nervy pilot punched through an evil veil of mist no more than 20 feet above the forest canopy. Suddenly we were looking out over the open grasslands of the Baiyer-Wahgi Divide. What a collective sigh of relief inside that helicopter! The machine swooped down this open valley toward the welcoming Kagamuga strip—our gateway to Western civilization.

From the safety of the Air Niugini boarding lounge, Andrew and I could shrug off the helicopter ride as an aberration and instead reflect on a great month of fieldwork in the remote central highlands of Papua New Guinea. We had teamed up with a mammalogist from the University of Adelaide in South Australia, a herpetologist from the South Australian Museum, a botanist from Harvard University, and a team of local environmental officers from provincial and national departments, to make the first comprehensive biological survey of the Mount Kaijende highlands. This high plateau is studded with 12,000-foot peaks, tucked between the Porgera Valley (site of a giant gold mine) and the Tari Valley (home of the Huli wig-men).

The point of a biodiversity survey or biological reconnaissance is simply to enumerate the plants and animals living in the area of interest. In this instance, we were helping the mine's owner make a plan for offsetting the impact of the mine through a conservation set-aside. We were delineating the richest patch of habitat for conservation.

To study birds in this area I generally used five tools: mist-nets for trapping birds alive, a mini–disc recorder and a microphone to record birdsong, a pair of binoculars, and Papua New Guinean village naturalists to provide advice on identification and occurrence of local species. On this trip I used these tools to locate 102 species of birds at our four mountain survey sites, which ranged from 7,000 to 11,500 feet above sea level. Given that birds are the best-known creatures in New Guinea, I was really fine-tuning our knowledge of local bird distribution in the highlands of New Guinea. I recorded, for example, the first sighting in Enga Province of a rare high-elevation honeyeater, the Long-bearded Melidectes.

I also found that the lovely Ribbon-tailed Bird of Paradise inhabited the Kaijende area, but not its closest relative, Stephanie's Astrapia. In Stephanie's absence, the highland-dwelling Ribbon-tail expands its range down to 7,000 feet above sea level. It is always fun to learn something new about a rare bird of paradise, and this was the last of the group to be named—described to science in 1939.

Other members of the team used different tools, and made more momentous discoveries, in part because there had been less research on mammals, frogs, and plants. Our mammalogist, Kris Helgen, discovered a new species of miniature bandicoot—a terrestrial mammal something like a marsupial hedgehog. Herpetologist Steve Richards discovered six new species of frogs. Wayne Takeuchi, the botanist, collected some five hundred species of plants, fifteen of them new to science.

For a field biologist, nothing is remotely as exciting as being in a new place where there may be undiscovered species. Although our field trip in November 2005 to the Foja Mountains of far western New Guinea generated many more discoveries and much greater publicity (Chapter 10), the Mount Kaijende expedition produced its fair share of remarkable unknowns. Identifying a new species is a bit like establishing an athletic record, one that cannot subsequently be broken. The scientist who describes a new species gets to give the species its name, and the scientist's own name is forever allied to that new species—a little piece of immortality. And it is not just finding new species. It is learning the song of a bird for the first time, seeing for the first time some bizarre animal behavior, admiring a rainbow over a lost lake where no scientist has ever camped, being out at the very edge. It is being alive in every sense.

CHAPTER ONE

In the Rainforest

We went along in single file, and in silence, our men

being too heavily laden to chatter. The forest was

dense, the air moist and refreshing; all was silence . . .

we occasionally heard the cooing of a pigeon and the

hoarse cry of the megapodio.

Luigi M. D'Albertis,

New Guinea: What I Did and What I Saw (1880)

NEW GUINEA—LARGER THAN MADAGASCAR or Borneo—is amazing for its plant life (more than fifteen thousand species), its bird life (more than seven hundred species), and its human diversity (more than a thousand distinct language groups inhabit this great rugged island). New Guinea has expanses of lowland jungle and a great Andes-like mountain chain that rises in places to more than 15,000 feet. (Throughout this book, the name New Guinea is applied as a nonpolitical, geographic term, to indicate the world's second largest island. By contrast, Papua New Guinea, or PNG, is the island nation that comprises the eastern half of the island of New Guinea as well as additional island groups to the north and east. I use the political term Papua, formerly Irian Jaya, to indicate the western half of New Guinea, which is a province of Indonesia.)

New Guinea is sometimes overlooked by naturalists and wildlife enthusiasts because its mammal fauna cannot hold a candle to those of Africa or India or South America. Geologically one with Australia, the biota of the two have the same origins. Thus New Guinea has many kinds of marsupials, and all the world's species of the egg-laying monotremes except for the platypus. Like Australia, New Guinea lacks native species of squirrels, deer, cats, canines, monkeys, and other large placental mammals; the two regions have only rodents and bats. New Guinea's mammalian claim to fame is its tree-kangaroos and its giant rats. Nice, but not mind-blowing. Hence, New Guinea, wonderful for an ornithologist, botanist, and bat expert, is rather impoverished in the kind of big-animal fauna known elsewhere.

I have been working in New Guinea since 1975, mainly because of the allure of the birds of paradise, the bowerbirds, the honeyeaters, the cassowaries, and the other bird families unique to the region—weird and wonderful birds galore. The idea of going to New Guinea came from my father's association with Tom Gilliard. As boys, both had been members of the Ruxton Naturalists Club in suburban Baltimore, collecting bird's eggs and snakes, doing the things boys like to do. Gilliard grew up to become the world's expert on the birds of New Guinea, especially the wonderfully diverse birds of paradise. As a boy, I pored over a big picture book of birds of the world authored by Gilliard, and I began to dream of adventures in search of rare tropical birds.

Opposite: Sulphur-crested Cockatoos

As a senior in college I won a Watson Fellowship, which financed a year in New Guinea to study the birds of paradise. That first sojourn changed my life. Now, after more than forty trips there and more than six years spent living in New Guinea, I cannot imagine a year going by without a visit to this far corner of the tropics.

⋙⋘

Have you ever watched one of those shows on Public Broadcasting that features a naturalist doing something in the jungle, such as studying the Orangutan in Borneo? If so, you might have wondered how that naturalist ever got the nerve to go out and camp in the middle of the rainforest, especially for the first time. Let me try to answer that question by telling the story of my first jungle camp in Papua New Guinea. In doing so, I will try to show how every journey into the rainforest teaches lessons that are hard to come by any other way.

On September 24, 1975, I established my first jungle camp in a corner of the rugged valley of the Upper Watut River. I was prudently following a path forged by several esteemed predecessors. This area had been visited by the field ornithologist Herbert Stevens in the early 1930s, when he worked out of rough mining camps with names like Surprise Creek and Bulowat. At the height of the Morobe Province gold rush, the crude but available infrastructure built up by prospectors made it possible for the adventurous Stevens to get into New Guinea's forbidding interior to collect birds. In the winter of 1974–75 I had spent several months at Harvard's Museum of Comparative Zoology studying Stevens' bird collection from this area, teaching myself the birds of Morobe Province. Stevens' study skins are among the finest ever brought back from New Guinea—perfectly prepared and labeled with details on each bird, the labels written in a small but neat script. In 1974 there were no illustrated guides to the birds or mammals, so studying museum specimens was the only way to learn New Guinea's wildlife in advance of going there.

In 1967 mammalogists Abid beg Mirza and Allan Ziegler surveyed birds and mammals in a side valley of the Upper Watut. They, like Stevens, generated a fine collection of the region's birds and mammals. These are the building blocks of field zoology. I had pored over their collection at the Bishop Museum in Honolulu and also at the small zoological collection at the Wau Ecology Institute, my New Guinea base of operation. Stevens, Mirza, and Ziegler had "struck gold" in the Upper Watut, and I wanted to get to that golden spot. I was curious to see how things had changed since their visit.

I was also particularly eager to locate a communal display tree of the Raggiana Bird of Paradise, Papua New Guinea's national bird, and perhaps the most beautiful of this family of gorgeously plumed species. I had glimpsed the Raggiana before in Wau, but had never had the luck to experience the otherworldly sight and sound of this species in full group display. Like the display of the Cock-of-the-Rock in Suriname, it is one of the pinnacles of avian evolution and mating behavior.

Back then I was twenty-three years old, an untested and self-styled student of tropical ecology on a Watson Fellowship year abroad. I headed to the jungle with two Pidgin-speaking Papua New Guinean guides, a venture that caused me no little trepidation.

Because he did not know English, Tawi Bukum, the driver, and I could only converse in Pidgin, the common trade language in Papua New Guinea at the time. Stumbling over the words, I told him to try to locate the creek crossing where the old dirt logging road passed nearest the Ziegler research campsite: *"Tawi, gutpela pren, yumi mas painim dispela ass bilong rot igo insait long bik bus na kamap long olpela kamp bilong tupela masta olikalim Abid na Allan."* He nodded mutely.

During the hour-long drive from my base of operations in the town of Wau, in the interior hilly heartland of Morobe Province, I had a case of pre-expedition butterflies. Was I ready to lead a field trip into unknown bush in a Papua New Guinea that had gained political independence from Australia only two weeks before?

The sun stood high in the cloudless, pale blue sky when Tawi stopped the jeep on the dirt track by a sagging log bridge. (Loggers do not build for the long term; their unwritten code is "Get the logs out, then get the hell out.") Tawi told us, *"Bai yupela behainim dispela liklik wara igo antap"*—

Pidgin for "Follow this creek upstream." That was all we were to learn today from the ever-taciturn Mr. Bukum. Any sign of a logging track was long gone after eight years of rapid tropical regrowth under the bright Watut sun and abundant Watut rain.

Titi and Timis, my guides, unloaded our packbags from the jeep and we struggled to hoist them onto our backs. We had ten days' food, as well as the necessary camp and research supplies, strapped onto World War II–vintage plywood military packframes (their date of fabrication, "1944," was stenciled on the side of each). Our loads included plastic sheeting for a camp roof, binoculars, mist-nets, tin pots, enameled metal plates and cups, and food. We had to tote it all to a site suitable to make camp. How far, we did not know at this point. It was hot and I was already sweating profusely. Although the butterflies in my stomach were gone, an actual swarm of small white *Delias* butterflies boiled up like animated apple blossoms from the wet orange clay by the rushing stream that constituted our gateway into the forest.

As no path was evident, we plunged into grass that was hip high, feeling our way with our feet—bare feet in the case of Titi and Timis. My feet were booted, but my legs were bare, and the guides neglected to point out the stinging nettles that infested the river edge. Pain lashed my calves and I found myself cursing and dancing about to avoid the stinging leaves. I spouted a fractured sentence in pidgin, abusing my assistants for their insensitivity to my white skin's tender vulnerability. Titi and Timis laughed loudly. In Papua New Guinea another's misfortune (especially in the form of physical pain) is reason for amusement.

We crabbed our way upstream for a very slow several hundred feet (every plant a potential nettle to my fevered eyes) before entering the deep shade of the forest interior. Here a foot trace immediately became visible, leading us on. My stinging legs were forgotten for a moment as I gazed upward into the forest canopy. It was my first step into tropical ravine-bottom forest—tall, heavily shaded, and tangled with vines. Jungle! Once in the forest interior we continued to move upstream in fits and starts, looking for a flat place to camp. The ravine was deep and narrow, so more than an hour passed before we were able to locate a small spot above the stream that could serve as a place to sleep and work. We never did find the site of Ziegler and Mirza's original field camp.

We dropped our packs against a convenient tree, and Timis and Titi put their machetes (or bushknives) to work, cutting and slashing at every bit of low vegetation in a circle about 30 feet across, encompassing the flat patch and also including some rougher terrain. Titi and Timis were masters at thrashing a field camp into shape.

After the undergrowth and saplings were downed and removed, we turned to the more serious job of taking down the larger understory trees. The men demonstrated a neat trick that I would subsequently see used time and time again in PNG. Five or so small trees needed to go. Each was 3–5 inches in diameter and 15–30 feet tall. Instead of attacking each one singly, they planned to fell the group as one. Timis took our ax and Titi the larger bushknife, and they cut all the trunks most of the way through. They then cut at the very largest of the group in such a way that when it fell, it came crashing down onto the rest. Because of the vines and intermingled branches, the trees shared a common fate, and their demise created a hellacious din as they spun and twisted and tumbled off the edge of our miniature plateau. We hooted in appreciation, and an unknown bird sang out in response (a White-crowned Koel, a black cuckoo that I later learned calls when it hears loud noises in the forest). The trees fell, and a flurry of leaves rained upon us. A few rays of sunlight unexpectedly settled down onto our little patch. It was about 1 P.M., one of the few hours of the day when the sun would pierce the shadows of this jungle ravine bottom.

We had a lunchtime snack and quickly went back to work. By 2:30 we had a camp, and none too soon. After we had stretched the long strips of plastic sheeting over the lean-to's roof, the rain began to fall—hard and steady for about fifteen minutes. Titi coaxed a fire under an aluminum pot and we drank some tea and watched the post-shower drip from the canopy vegetation. Because of the drip, the effects of the shower lingered for a half hour longer than the actual event. The rainwater passes from plant to plant and only reaches the ground after the force of its fall has been broken once, twice, or more by the intercepting greenery.

Think of this wondrous aspect of the rainforest structure. Many leaves sharing each raindrop has multiple benefits: it disseminates the water to the various thirsty plants; it raises the interior humidity of the forest through repeated breakdown of each raindrop; it absorbs the kinetic energy imparted by each falling raindrop and thus protects the forest floor from erosion, keeping rainforest streams relatively free of silt. All of these benefits are lost when a rainforest is cleared. Luckily, the recent logging in our section of the Upper Watut had been highly selective; the company had taken the giant softwoods scattered through this largely hardwood forest. As a result, the character of the forest remained essentially intact.

❦❦

Having established camp, we needed to string up our mist-nets. Made of fine black nylon thread and strung between poles like an oversized badminton net, mist-nets catch birds without harming them. The ones I had were about 40 feet long and 9 feet tall, far bigger than any badminton net. Once set in the forest, each nearly invisible black net would act like an oversized baggy spiderweb, holding fast any birds that passed through the forest understory where the nets stood. The nets would hold the birds safely captive until we could extract them, identify them, band them, and release them.

Because we were in a ravine, finding appropriate places to locate the nets was difficult. Luckily, I had Titi and Timis. Although they spoke little English and I little Pidgin, we all knew the job at hand. My two companions had worked two years for Thane Pratt, another student bird researcher who had preceded me in Papua New Guinea. Thane had a real knack for bush lore and acted as a mentor to me during my first weeks in the country. Timis and Titi already had lots of experience with all the kinds of field activities I wanted to carry out over this fortnight.

The men knew just where to place the nets to capture the most birds. Unlike me, they had no fear of the unknown. To them this patch of bush was just like somebody's back yard. It was merely a matter of getting to know it in the first two days by clambering through the forest, following their noses, clearing the vines with a sharp bushknife. They would go out for several hours at a time, then return to tell me about the distant spots they had visited. They were my eyes and ears—my rainforest radar—for those first few days in the Upper Watut.

We had all of the nets up by the end of the third day. The virtue of nets is that they are always on the job; it's just a matter of checking them regularly. The nets were to produce the biggest ornithological surprises of our stay, for I was still a novice at bird-spotting in tropical forest. In New Guinea, many species are nearly impossible to see even when present at close range, perhaps because, with few large game mammals in the forest, birds have been the main prey of hunters for millennia, and only the most wary have survived. This vast island is a land of hunters of all things edible.

It is also the land of the bird of paradise. Of the thirty-eight species that inhabit the Australo-Papuan region, about ten live in and around the forest where we were camped. The species vary remarkably. For example, the diminutive King Bird of Paradise is bluebird-sized and glistening red and white. The Crinkle-collared Manucode is like a compact crow, shiny black with highlights of purple and blue. The Magnificent Riflebird is chunky, long billed, and nearly tailless, the male being glossy black with a iridescent blue throat shield, the female buff brown with fine barring underneath. The Manucode is monogamous, the Riflebird and King are polygamous. In 1975 little was known of the habits of these forest dwellers. Here was a research opportunity waiting to happen.

We were camped at 7 degrees, 8 minutes south latitude, and 146 degrees, 34 minutes east longitude—a few miles southwest of the hamlet of Latep, on the road to Otibanda, in a rugged interior upland valley about 35 miles from the coast, in Morobe Province. The camp stood at about 2,475 feet above sea level, with the surrounding hills rising steeply to more than 3,300 feet. In terms of global geography, we were in the middle of nowhere.

The three of us camped under a 12- by 15-foot roof of black plastic sheeting. We slept in blankets laid on plastic on the ground, and our only privacy was provided by the mosquito nets draped over each of our sleeping places.

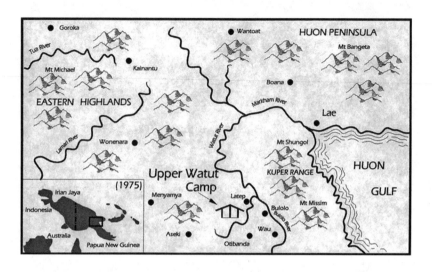

Our "field laboratory" consisted of a corner of the lean-to in which Titi had fashioned a rough table and bench made from saplings bound together by *busrop* (bush-rope)—pliable and tough vines stripped from trees. The men also fashioned an "outside" work area that was open to the elements—a larger sapling table and a bench with a back, overlooking the stream. That was it. The rest was provided by nature. We drank straight from the forest stream. Heating and air-conditioning were provided by the rising and setting of the sun in relation to the forest and the ravine topography.

The sun was a problem for us, in that we were camped at the bottom of a deep ravine. Whereas the sun rose at 6 A.M., in the ravine bottom the gloom of dawn extended to about 8 A.M. The continuous roaring of the stream made it difficult for me to rise early without the diligence of my two assistants. One of the highpoints of the day was the hour when the sun shone down onto our camp. Then I bathed happily in the cool stream and we aired out our sleeping bags and blankets. Until one has attempted to take a photograph with slow film, it is easy to forget that the interior of tropical humid forests is dark, especially in a deep forested valley.

Bathing in the cool stream was sweet leisure, because it was hot, sticky, dirty work we were doing. What a pleasure to settle into the little gravel-bottomed pool of a clear mountain stream and let the grime rinse away, while the sun played on the water hole. At these times the sun was a friend, not a foe.

One scourge in Papua New Guinea's jungle was the little red chigger mites. In search of a blood meal, they crawled into our clothing and burrowed into vulnerable patches of tender skin. Overnight these became pink swellings, which itched for days. Most rainforest researchers fear chiggers more than any other pest.

Even more annoying were the sweat bees—so small, so harmless, so friendly, and yet so incredibly pesky. A swarm of these brownish creatures (each about a quarter of an inch long) would descend on camp during the middle of the day, in search of salt, sweat, sugar, whatever. They would alight on anything and everything and taste every surface. They would not be denied. As I sat on the bench by the stream, hundreds of these little critters would find their way onto my short-sleeved shirt, onto my arms and neck, and into my scalp. Once there, they would begin searching and licking. Their tiny feet would tickle my skin. Their little "tasters" would prickle. They particularly liked to get tangled in the hair of my forearms. With their harmless persistence and vast numbers, they were impossible to ignore.

We were lucky on one count: the site had few leeches. These little leaf-dwelling bloodsuckers inhabit some of the wetter lowland sites in abundance, but New Guinea's leeches are insignificant when compared to those of the wet forests of Sri Lanka and northeastern India. Also, the mosquitoes were troublesome only after dark, by which time we had retreated under our sleeping nets.

We cooked over a small fire at the edge of our lean-to shelter, and we dined sitting on the bench, watching the stream. The meals were simple: a breakfast of hard biscuits, hot oatmeal, tea; a lunch of hard biscuits with peanut butter and perhaps a cookie; a dinner of stew and rice, finished with a cup of hot tea with the smoky flavor of the burning fire. Dinner was clearly the main meal of the day. Hunger stoked by long hours of strenuous physical activity outdoors creates a remarkable appreciation for any kind of hot food in quantity, but I could never compete with my guides when it came to consumption. They would cook a large pot of white rice. Each would heap his plate to overflowing, adding just a garnish of the "stew" that colored the crest of the rice mound. They would then wolf it all down.

⋙⋘

The men took charge of most of the domestic chores, so that I was able to concentrate on what I had come for—the wildlife. Except that I needed the men to assist me there, as well. I was in a fairly tough spot for a naturalist. I did not know the country. I did not know the birdsongs. I had difficulty seeing the birds. And I was a bit afraid of going out into the trailless forest on my own because I did not want to get lost. I started slowly, by wandering around the ravine bottom in search of spots where I could catch a glimpse of something. Although the undergrowth was not very thick, it seemed that most of the birds were in the canopy vegetation. I could hear them singing, but it was very difficult to see them well enough to identify them.

Of course, I quickly encountered common species I knew from Wau. The all-black Spangled Drongo was there, erectly perched, flicking its "fish tail" and emitting a series of metallic and musical notes. This drongo is a classic insect-catcher, perching and watching, then suddenly launching out after a flying insect—often one flushed up by another foraging bird. Drongos were inveterate followers of mixed-species flocks. I have learned that if I listen for the songs of drongos, I can often locate a bustling mixed flock in the forest nearby.

The Hooded Pitohui was another common forest bird. It looks a bit like a slightly oversized Orchard Oriole, with a black head, wings, and tail, and a chestnut brown body. It is not unusual to see small parties of Hooded Pitohuis in association with drongos. The pitohuis tend to skulk through the forest, in either the canopy or the understory, and are vocal but shy. They hunt for insects on branches and larger limbs, and visit trees that are carrying small fruits, especially figs.

We netted a Hooded Pitohui on our fifth day in the camp, and it elicited a curious response in the men. Titi and Timis had one difficulty with mist-netting birds, and that was that they hated to release the birds after capturing

The chemically defended Hooded Pitohui inhabits
only the hill forests of New Guinea.

them. Papua New Guineans consider birds and all other small vertebrates
as "game," so it seemed unnatural not to consume all that we captured. But
when I pulled the pitohui out of the cloth holding bag after weighing it, Titi
looked at it and grimaced. He said this was a bird that tasted bad—and thus
was one of the few animals not regularly consumed by his clan. At the time I
found this distinction mildly interesting and I noted it in my fieldbook, but
I never pursued the issue.

More than a decade later, Jack Dumbacher, a research student working with me in the hills behind Port Moresby, discovered the secret of the
Hooded Pitohui. Dumbacher, with assistance from scientists at the U.S.
National Institutes of Health, was able to show that the pitohui's feathers
and skin contain a potent neurotoxic alkaloid called homobatrachotoxin—a
molecule identical to that in the skin of one of the poison-dart frogs of South
America. It was the first poisonous bird discovered by Western scientists. Of
course, Titi and his village had known all along that the bird was poisonous,
as did most hill communities in New Guinea that encountered the bird.

A key difference between Jack's experience with pitohuis and mine was
that he had been poisoned by them. After handling several of the birds, Jack
had touched his hand to his mouth, and before long his tongue and lips began
to tingle and become numb. That clue set Jack on the road to discovery. I

had handled dozens and dozens of pitohuis over more than a decade but never was zapped by their toxin, so Titi's warning did not tweak my scientific interest. The rest is history. In 2004 Jack showed that the Hooded Pitohui consumes a tiny Melyrid beetle that carries the toxin. Presumably the bird sequesters the beetle's toxin for its own chemical defense. Jack has postulated that the toxic defense of the pitohui is to protect it from predation by bird-eating snakes. The next step will be to locate the ultimate source of the toxin (called a steroidal alkaloid) that presumably feeds the beetle that feeds the bird. I suspect it is a plant or fungus.

Finding and identifying the poison's source in New Guinea will likely also tell us the source of the poison-dart frog's poison in South America. Many plants and fungi have ancient lineages that evolved before the breakup of Gondwana, the ancient southern supercontinent (South America+Antarctica +Australia+New Guinea). We presume that toxic plants of the same lineage would be living now in South America and New Guinea. In this way, the frog (known only from South America) and the bird (known only from New Guinea) are biologically linked by an ancient plant that has a now-disjunct widespread "Gondwanan" distribution.

After a few days, I found a small tree-fall clearing near our stream that allowed me to look up and down the ravine. I could perch there quietly and watch the birds moving through the upper levels of the forest, as well as the occasional individual that would come down to the stream to drink or bathe. Pairs or threesomes of the large, all-white Sulphur-crested Cockatoos sailed by, flapping their big rounded wings with their distinctly shallow wingbeats. If members of this species were not so annoyingly vocal and aggressive, they would be considered among the wonders of the bird world. Seeing them for the first time was spellbinding. I sat motionless and held my binoculars on these white apparitions of the forest canopy. Their powerful and unpleasant shrieks would break the peace when they saw me.

On one occasion, a single cockatoo perched high in one of the pine-like giant araucaria trees and snipped branchlets and dropped them down on me. The bird's magnificent recurved crest feathers were a pale yellow that was offset by its white cheek feathers, fanned up over the bill, and the pale blue skin around the eye. It was lovely to behold.

There is more to a Papuan rainforest than its birds, of course, although the birds are one of the most conspicuous elements. To an ornithologist focused on his particular quarry, the mammals almost do not exist, so rarely are they encountered. Aside from some persistent rats that raided our camp at night in search of food, I saw only a single free-ranging mammal in the entire two weeks. While perched on my log lookout in the late afternoon, my eye caught some movement in foliage high in a tree. With my binoculars I could just make out a rotund creature, mostly white and spotted with tan, moving slowly along a branch. Upon further inspection, I could see it was a cuscus (a chubby possum) with a prehensile tail and a pair of goggling eyes set on its round head—a Spotted Cuscus, one of the more common members of the Phalangeridae family.

Looking like a fuzzy ball of pale fur, it is entirely unlike any creature we see in the United States. Compared to our Virginia Opossum, the cuscus is larger, more thickly furred with short woolly hair, heftier, and almost exclusively arboreal. In contrast to the tiny eyes of our possum, the Spotted Cuscus has large, bulging, honey-brown eyes. It does not play possum and can be quite aggressive when defending itself. Its sharp teeth, powerfully clawed feet, pungent scent, and generally nasty disposition make it a mammal of uncertain charm. It is a remarkable sight when it is moving slothlike through a forest canopy, one of the few larger forest mammals that can be found in the daytime.

Still, in nearly thirty years I have seen this common species only five times in the wild. That typifies the cryptic nature of the forest mammals of New Guinea (as well as the general inability of ornithologists to find mammals). When I mentioned my mammal sighting to the two guides, they grew quite excited and not a little annoyed that I had not brought it to their attention immediately. They wanted the cuscus for an after-dinner snack. Everything edible in the forest is eaten by the local people.

＊＊

In New Guinea's hill forests, snakes are more in evidence than mammals, but they still are not terribly obvious. I had been told, incorrectly, that there were no poisonous snakes in the area. Titi and Timis saw no distinction between poisonous and nonpoisonous: all snakes were to be feared and killed, or at least avoided. Even the small, harmless forms were subjected to this attitude (not so different from the typical frame of mind in the United States). Because I totally lacked snake identification skills, I too could not afford to take chances.

One day I found a slim russet-brown snake entangled in one of my mist-nets. It was attempting to swallow a small bird. Not surprisingly, it was unable to do anything more than get itself royally tangled in the nylon netting. I removed it and photographed it. After the trip it was identified as a Brown Tree Snake. I have handled the species several times since then, and the smaller individuals are fairly docile. The venomous rear fangs seem not to pose any real threat to humans. But the 1980s were to see this very species become the culprit in a remarkable ecological drama played out on Guam, in the central Pacific. The snake apparently was accidentally introduced there during World War II, and after inconspicuously subsisting on the island for decades, its population suddenly exploded. Since the snake is a specialized bird predator, this expansion threatened the existence of virtually all of Guam's native birds, including the endemic and flightless Guam Rail and the Micronesian Kingfisher.

Only the intervention of field biologists and conservationists prevented the extinction of these birds. (They are now bred in captivity but apparently are gone from the remnant forests of Guam.) Millions of dollars have been invested by the U.S. government to control the snake and rescue the birds. There is now a realistic fear that the snake will invade Hawaii by way of the military planes that shuttle to and from Guam. This scenario could doom the few native bird species that remain in Hawaii's remnant upland forests. The Brown Tree Snake, the ultimate bird predator, is probably the aggressor that the Hooded Pitohui is seeking to avoid with its poison shield.

The Small-eyed Snake is a common and highly venomous lowland forest species from New Guinea

I saw only one free-ranging snake during our stay in this tributary of the Upper Watut, and it was a venomous species. During a late-afternoon walk along one of the old logging tracks that scarred the hills above our camp, I came upon a 4-foot-long snake lying motionless on a bare patch of earth just off the track. It was strikingly patterned in a manner that might be termed either hideous or strangely beautiful. Certainly it was unforgettable. The snake was buff tan with a ghastly pink sheen, further distinguished by a series of indistinct but ominous-looking darker bands around its dorsal midsection. Its small head was adorned with a pair of beady blackish eyes, hence the name Small-eyed Snake. It is one of the few venomous front-fanged snakes that inhabit the interior upland forests of New Guinea. A nip from it would prove unpleasant indeed. Fortunately, the species tends to be fairly docile. The individual I encountered made no hostile moves in my direction. Still, to a barefooted villager, the threat of treading on one of these snakes while it is sleeping on the smooth hard earth of a local footpath is real indeed. The villagers take no chances. They kill every snake they see.

One morning, huffing and puffing up a steep slope to our line of mist-nets, I could see a colorful apparition in the bottom of one of the nets. I had no idea what it was, but I knew it had to be something special. On closer inspection, I found one of earth's most beautiful birds, the Red-breasted Paradise-Kingfisher. Wings, crown, back, and tail range from deep blue to cerulean. Underparts vary from pink to salmon to deep orange-red. The bill is carmine. The twin central tail feathers are pale blue, greatly elongated and tipped with white. When perched, the bird waves these feathers slowly up and down. This kingfisher is a compact jewel, a third the size and bulk of North America's Belted Kingfisher, and vanishingly inconspicuous in the forest. Still, it is an ornithological treasure at close range.

Seeing this bird in the net was a shock and a delight, for Mirza and Ziegler had missed this species in their 1967 survey. I gently disentangled it from the net, placed it carefully in a cloth holding bag and tied it shut, then laid it in the upper portion of the net for safekeeping while I went to check the rest of the nets. Returning five minutes later, I found the cloth bag on the ground and my fairy jewel gone before I had had a chance to photograph and measure it. I was crushed.

That same afternoon I labored back up the hill to check the nets, and there was yet another surprise: a Shovel-billed Kingfisher—a big, weird, unlovely brute that is notable for its elusiveness, not its beauty. Patterned dark brown dorsally and buffy-brown below, with a pale collar, white throat, and turquoise rump, the bird is further differentiated by its broad and very abbreviated bill, which is unlike that of any other kingfisher. It is more a miniature version of the bill of a Shoebill Stork or a Hammerkop and is used mainly for digging earthworms from the forest floor.

The Shovel-bill struggled to escape the net when it saw me approach. I dropped to my knees and grabbed at it so it would not get away. It grabbed me simultaneously, taking my index finger in its bill and shaking it like a bulldog. Because it was quite painful, I hastened to pull out my finger, to no avail. The bird had a grip I could not pull away from, and it was not relenting. I got my other hand on the tip of the bill and pried its beak apart just enough to allow my finger to slip out. As I released the bird in order to tend to my pain, the kingfisher, with a stroke of its powerful wings, vaulted out of the net and flopped onto the ground, free. I lunged for it, forgetting the pain it had just visited on me, but the fugitive bird sped away, only to be ensnared in one of the other nets down the ridge. Breathlessly I ran down and got a firm grip on the brute's head (with a bandanna as protection) and wrestled it into a holding bag. One out of two is a fair return in the Papua New Guinean rainforest.

One disappointment of my mist-netting was that I was not catching any birds of paradise, which tended to remain in the canopy. My guides did point out several vocalizations of this bird they term *kumul.* I was getting more and more desperate to find a display tree of the Raggiana, and I reminded the men that this was a priority for me. They assured me that they would keep alert for one on their forest rambles.

Although the birds of paradise were not cooperating here in the Upper Watut, the forest butterflies tended to be a friendly and ever-present adornment of the forest. They brought the forest to life, especially during the slowest hours of the day. The sun and heat seeped down into the forest interior, warming the air, which then sucked up the abundant free moisture, creating Nature's own steam bath. Ghost-like owl butterflies flopped clumsily in the understory, seemingly obvious to the observer, but never allowing a close approach—white apparitions with staring eyes painted on their hind wings. (Do the "eyes" really reduce predation by birds?) The lovely morpho-like Blue Emperor Swallowtail darted rapidly up in the higher stages of the vegetation, flashing iridescent blue, then practically disappeared as its wings closed to show only its sooty-gray underwing colors. Little white *Delias* species, with their intricate and multicolor underwing patterns, tended to follow the stream in fluttery bands. Finally, the queen of the forest, the pied female Poseidon Birdwing, drifted on its 6-inch wings high above, maneuvering elegantly in breezes that never reached the ground at midday.

I would be roused from such a reverie by the rude and pungent aroma of *smok brus* (Smoke Bruce)—the incongruous name for the harsh tobacco grown in rural gardens and sold in village markets around Papua New Guinea. The smell of brus signaled the arrival of one of my New Guinean colleagues.

Both Titi and Timis smoked a lot, fashioning cigarettes from a pinch of brus rolled tightly in a torn rectangle of newspaper. This pastime was important to the men, as illustrated by their statement when they had run out of tobacco: *"mipela dai long simok"* (we are dying for lack of tobacco).

After about a week at our Upper Watut camp, the tobacco did run out. Timis, the heavier user, mumbled something to me about his need and I paid it no mind. When I next looked up, he was nowhere to be seen. Titi informed me that Timis had departed in search of smoke. But where? Where was he going to find some tobacco in this neck of the woods? I could not believe my ears. Then again, I was not terribly familiar with what was and was not possible in rural Papua New Guinea. Three hours later, Timis was back, glistening with sweat, bearing a big bundle of homegrown tobacco leaves. It happened that Titi and Timis had wandered far and wide while scoping out the area over the week we had been there. In doing so, they had traversed logging roads and actually bumped into some hunters. Timis had learned from them the location of the nearest village of Watut (Kukukuku) people on a ridge high above us. He had journeyed there to get his tobacco. All for a habit—a habit that was, and continues to be, nearly universal in rural Papua New Guinea. Smoking begins early and continues into old age.

Long yia 1975, olegeta manmeri husat bin istap insait long Morobe Province emi bin save gut long Tok Pisin. Tok Pisin emi winim olegeta kain tokples estap long Papua Niugini. Husat man i laik wok insait long bus insait long Papua Niugini, emi mas lainim dispela kain tok. Sapos nogat, hau bai emi toktok long ol? (In 1975, everyone who lived in Morobe Province knew how to speak *Tok Pisin* [Neomelanesian Pidgin]. Tok Pisin is the most popular language and is understood by more people than any other language, especially the local languages, in Papua New Guinea. Anyone who wishes to work in Papua New Guinea's backcountry must learn Tok Pisin, otherwise he won't be able to make himself understood.)

Neomelanesian Pidgin is a trade language that is virtually universal in Papua New Guinea, although many more people speak English well today than in the 1970s, when Pidgin was dominant in the PNG heartland—the northern parts of the interior. Part of working in the field is learning to speak and understand Pidgin. And the best way to do so is to go into the bush with assistants whose only means of communication with the field scientist is Pidgin.

For the locals, Pidgin is a second or third language, for it is no one's birth language. In village Papua New Guinea, everyone has a local language (known in Pidgin as a *Tok Ples*—talk place), and this language of course has primacy over any other. More than eight hundred local language groups exist in Papua New Guinea, many of which are spoken by only a few people, Pidgin is used to communicate with those who do not share a common Tok Ples. Without Pidgin (and its southern counterpart, *Motu*) Papua New Guinea in the 1970s would have been hopelessly fragmented, culturally and socially. Today English is the language of choice, although Pidgin is spoken in informal settings by everyone, everywhere.

That Pidgin is a product of past colonialism quickly becomes clear to a beginning speaker. A young child is called (even today!) a *pikinini,* whereas a slightly older child is called a *manki* (monkey). In 1975 a white man was commonly referred to as a *masta,* and a white woman was a *missis;* a Papua New Guinean man was a *boi* and woman was a *meri.* These terms, thank goodness, have died out.

Other Pidgin usages are amusing without (I think) being offensive. In this almost universally Christian country, Christ is *Pikinini bilong God.* A beard is *maus gras* (mouth grass). Colorful expressions abound. Make no mistake, however: I am poking light fun at the language, but it is a real language—an important and functional working language, though one mainly spoken. It is less useful for extended exposition.

Timis and Titi, Pidgin speaking, chain smoking, and barefoot, are among the very finest naturalists and woodsmen I have ever known. For me, a novice naturalist in Papua New Guinea, these two men (*tupela wokman*) served as my field identification guide, my trail map, my eyes, and my ears in the field. Anything I could do, they could do better.

Because of their daily exertions, the average village-raised Papua New

Guinea man is strong, patient, hardworking, and persistent. Aside from sheer physical skills, many of the rural-living men are experts at identification of the plants and animals, which frequently had species-specific names in the local language (but lacking in Pidgin). Once my guides were able to match the scientific name of a bird to their own local name for it, they could effortlessly provide me with identifications. The contribution of these barefoot naturalists did not end with mere woodcraft skills and field identification; they also had accounts of the habits and life histories of the creatures of the forest.

Both Timis and Titi grew up in villages situated in mountainous terrain not unlike the Upper Watut. They walked the tiny hard-packed traces that served as *rot bilong wokabaut* (roads for foot travel), and they daily crossed stream gorges, down one side and up the other. Our work on the Upper Watut was a lark by comparison, because they were not carrying water or firewood to ridgetop homesteads, or lugging a 60-pound cassowary back to the house for a feast. We were in their environment and they prospered.

For me it was another story. My legs were always tired. I wanted to park myself on the log by the stream, but I knew I had to climb up out of this valley bottom to search for birds. My ankles were sore, my knees were sore—it was basic training Papua New Guinea–style. But the forest was new to me and I had to see as much of it as I could. The incentive was new birds. Virtually every time I went out, I saw something unfamiliar. And the forest held other wonders: giant trees, huge epiphytic orchids, speedy monitor lizards. It was *all* new to me.

Looking back on this trip from the perspective of thirty years, I can see my mistakes. First, we camped beside a rushing mountain stream. That was fine for access to drinking water and for bathing, but the sound of the tumbling stream drowned out all but the loudest birdsongs. It is a nice sound for sleeping, but poor for bird inventory. Second, we camped at the bottom of a 500-foot ravine, so there was nowhere to go but up. Every day we had to labor uphill in search of birds. I am not talking about some gradual incline, I am talking ravine face. In some places we had to cut steps in the ground to get up and down. It was tough to reach observation sites by dawn, when the birds are most active.

Another thing I failed to do was to obtain permission to camp from the traditional landowners of the forest. I was under the mistaken impression that this forest was a no-man's-land and that anybody could visit or use it. I would learn over time that all of Papua New Guinea's land is customarily owned by tribal people scattered about the landscape. What looked like "wilderness" to my Western eyes was somebody's backyard or their hunting or gardening territory. It was an important lesson, and it figures prominently in the work I do today fostering nature conservation in the southwest Pacific. In any event, our little camp was so isolated that we never encountered another soul. In some parts of Papua New Guinea, trespassing is a serious offense, with justice meted out by the angry landowners themselves.

The araucaria pines sprinkled throughout this rugged landscape of the Upper Watut were of intense interest to the Commonwealth New Guinea Timbers Company in downtown Bulolo, some ten miles from our camp. In fact, the wonderfully tall araucarias I was encountering in the forest around our camp were but a remnant of the area's great stands of this ancient gymnosperm. By the time we arrived, the loggers had already come in and "high-graded" the forest—removed the most valuable timber, primarily the largest araucarias. I should not have been surprised: Bulolo was a major timber-processing site, precisely because of the presence of these giant conifers. The extent of the extraction only became evident to me after I wandered out of our little unlogged ravine with Titi and Timis. Each ridge-crest bore a main timber access road, and each had small spur roads that led to the most important stands of trees.

The first time I hiked to the nearest ridge-top was a shock. I climbed through superb old-growth forest on the steep ravine sides and stepped out onto the sun-baked orange clay of the timber road, already badly eroded by the rains. It formed an orange gash through the heart of this hill forest. I was initially pleased to have access to a ridge-top roadbed for making observations of birds. It was, after all, an almost ideal pathway from which to bird-watch—a far cry from my streamside perch on the log in the ravine bottom!

But the more I looked around, the more dismayed I became by what I saw: old pieces of rotting logs, some of remarkable girth; the remains of forest giants now gone, the aftermath of selective logging. Much of the original forest still stood. Yet the evidence of needless waste was everywhere. For every large tree taken, it appeared that at least a dozen medium-sized trees had been sacrificed. At every place the sun penetrated, it was dry and scrubby, the openings now choking with weedy growth.

Standing on the eroding bed of the main logging road, with Black-capped Lories noisily racing by, the men quietly told me they had a surprise for me. They would not say what it was as they guided me to a secondary logging track that led down onto another forested ridge. Here the hill forest was grand, with giant strangler figs and a bewildering variety of hardwood species. As we carefully picked our way down the track, I began to hear a periodic crescendo of raucous bird calls. A minute later we were standing beneath a huge mahogany tree with large spreading limbs. About a hundred feet up I could see five glowing orange forms moving frantically on a cluster of small branches. Every minute or so the sound would swell as these Raggiana Birds of Paradise shouted for attention: *wa wa wa WAH WAH WAH WAU WAU WAU WAU . . . WAU . . . WAU.* Focusing my binoculars on the assembly, I could see several adult male birds that were gorgeously plumed—orange, green, yellow, tan, brown, blackish-brown—and agitated. The males shook their long flaming filamentous flank plumes in excitement. They were being shadowed by several other young males in various stages of plumage development. Finally, I spotted two rather skittish females cloaked in their drab brown and tan. This was a communal mating display, in which the males danced to attract the choosy and shy females for mating. It was something like an avian bordello in reverse: the males dressed up and hung out, and the females came for furtive visits from time to time.

I spent the rest of the day lying on my back on the forest floor, gazing up at these birds doing their thing. When females were present, the action soared and there was singing, flapping, posturing, chasing, tumbling, pecking. When the females disappeared, there was desultory cawing, preening, and quiet perching, the occasional adult male pulling leaves to clear them from the display area. I had, at long last, witnessed the lek mating display of the Raggiana—the rare communal display that is employed by Sage Grouse in Wyoming, by Uganda Kob in Africa, by manakins in Panama, even by some dragonflies. My trip was now complete. I could go home satisfied that I had experienced the most extraordinary mating behavior in the bird world.

On the appointed morning we broke camp and carried our considerably lighter packs down to the roadhead. Tawi picked us up and we drove the rutted dirt road down the Upper Watut drainage toward the Bulolo Valley and then turned onto the wide gravel main road that followed the Bulolo River upstream to Wau. On this drive I noticed how humans had altered the natural environment in this rather remote corner of the earth. As we descended, it was easy to see that the forest on either side of the road had been badly hacked up by commercial logging, and in some places nothing but scrub remained.

As we crossed into the lower Bulolo Valley, the hilly countryside was mostly grasslands, the result of generations of subsistence gardening and seasonal fires set by traditional Buang or Biangai hunters and, in the 1970s, cattle grazing by colonial plantation owners. In the middle and upper Bulolo I found near-total devastation. The action of large-scale alluvial gold dredging had turned over the entire (once forested) floodplain of the valley floor to some considerable depth, to get at the gold in the gravels below. The nearby hills were an unnatural dark green, where monoculture plantations of araucaria had been planted a decade before. A fire burned in a patch of dry grass by the road. Squatter huts clustered in small settlements near artificial ponds created by the dredging of decades past. The squatters walked along the roadside with poles for fishing for the tilapia fish that had been introduced to the valley.

As we neared the gorge of the Bulolo River, which wends its way up to the Wau Valley, off to the right was a flat patch of old lakebed that had escaped the dredging. The few giant tropical hardwoods that still stood in this cattle pasture were evidence of the majestic forest that had once dominated this beautiful site. Not a single acre of the valley floor had been set aside for study or for regeneration. Remarkable, considering that Bulolo is home to Papua New Guinea's forestry college! Why is it that forest industries, when given the opportunity, save not one acre of prime forest for the future? The only forest reserve in the Bulolo Valley is Mount Susu (Breast Mountain), two steep-sided hillocks behind the town. Here a reserve of some two hundred acres has been set aside, where a few dozen old araucarias cling to the side of the rocky hill where, I presume, it was too steep to log.

After entering the rugged Bulolo gorge we came to a sign that announced MacAdam National Park. The far side of the gorge supported a superb stand of Hoop Pine (*Araucaria cunninghamii*) that was inaccessible to the loggers. Virtually every araucaria, and virtually every other forest tree, had been stripped from the side of the gorge where the automobile road stood. A narrow strip of ravine-side forest had been set aside as this national park. In the late 1980s the park was invaded by squatters and gardeners and miners, then burned over during a severe drought in 1997. Apparently none of the original forest remains.

On this, my first solo field trip, I learned little more than the basics of bush lore and field observation in the rainforest. In retrospect, the big potential discovery could have been the poisonous nature of the pitohui, but being new to this environment, I simply did not know enough to filter the important from the unimportant detail. I also did not yet recognize the importance of local informants to my research, so I failed to focus on what Titi had told me that day.

The main reason I made camp in the Watut was to learn the hill-forest bird fauna. I had started my fellowship years in New Guinea studying mountain birds, and this was my first step toward learning the bird life of lower elevations. I would eventually conduct bird censuses from the 11,000-foot summits of Morobe Province to the coastline, learning the birds from top to bottom. This knowledge ultimately led to my collaboration with Thane Pratt and Dale Zimmerman in 1986 on a comprehensive field guide to the birds of New Guinea. Even though the results of this first expedition did not yield scientific discoveries, it laid the necessary foundation of knowledge that I would use repeatedly later in my career. A lesson learned: Everything matters over the long run.

All knowledge has context. Much of what I had learned in the United States I had to set aside in this new environment. My first months in Papua New Guinea would find me assimilating a whole new knowledge base—life skills that would help me in the jungle. I learned that those who walk the forest barefoot can teach me a great deal. In 2005, when I most recently ventured into the forest in Papua New Guinea, I went with a local naturalist who taught me the birdsongs of his local landscape. In forests as rich as these, one's education never stops.

My second lesson concerned the Western illusion of remoteness and the inexorable reach of humankind. On the world map, or if I run Google "Earth," my little camp on the Upper Watut appears remote and isolated and forbidding. But humans are scattered all over our planet, and every community has been harvesting, and cutting, and exploiting the resources around it for longer than we care to think. The jungles where we camp and study are usually heavily impacted in ways that are not immediately obvious to the visiting outsider. We may notice the evidence of logging, but more important, subtle impacts have been caused by centuries of traditional use. It is difficult to get to the rare places where the impact of humankind is entirely absent. Even locations that today are free of people often suffered human exploitation at an earlier time. Virtually every forest on the island of New Guinea is human altered or human managed. It took me more than a decade to learn this lesson, and to begin looking more closely at the long and significant relationship between humans and forests there. This knowledge makes the truly pristine, unvisited sites (see Chapter 10) all the more remarkable.

CHAPTER TWO

In the Zone and on the Plantation

I went for a walk with Tom Gilliard, who was

working as Dr. Chapman's assistant. Both of us were

new to the inhabitants of the rain forest of Central

America, so when a large boar peccary appeared in

the trail, advancing towards us with a determined

gait, we glanced at each other. "Isn't this the point

where one climbs a tree?" I asked.

Dillon Ripley,

Trail of the Money Bird (1942)

AFTER FIFTEEN MONTHS IN PAPUA NEW GUINEA, I returned to the United States in July 1976 to begin graduate studies in ecology at Princeton University. In my second year I took a tropical field ecology course, held in Panama during January 1978. It was taught by John Terborgh, my research supervisor, who expected all his graduate students to participate. But this was no hardship. I was anxious to get my first taste of a neotropical rainforest. In Panama one awakens before dawn to the roar of the Black Howler Monkey, and in some ways the diverse and colorful jungle birds take a back seat to the mammals, especially the primates.

I had never seen a primate in nature. After hearing the howler monkeys the first morning in the field, I was eager to have a look at one. But my wish list was larger. I wanted to see a jungle cat in the wild as well—encountering an Ocelot or a Jaguar in the forest was an exciting prospect. I knew my chances were slim, given the way wild cats are treated around the world (pelts, coats, animal exhibits). Yet I wanted to immerse myself in the lore of the Neotropics, and cats are so emblematic of the jungles of the New World.

❦

Panama in 1978 was a friendly and chaotic place, quite to my liking as a brash second-year graduate student with not a worry in the world. The airport at Tocumen seemed remarkably dilapidated and small, and the birds just outside the door of the terminal seemed wonderful (Great-tailed Grackles, Tropical Kingbirds, Fork-tailed Flycatchers). I chattered in broken Spanish to the cheerful taxi driver as he ferried the eleven of us to our one-star hotel in Panama City, where we stayed while preparing for our campout in the Canal Zone, that now-deceased slip of United States–owned territory cutting across Panama's midsection. Panama City was crowded and bustling and a far cry from the only other tropical city I had known—dusty little Port Moresby in Papua New Guinea. I could see big office towers, huge banks, large boulevards choked with cars.

Looking back, I remember best the food and the cold beer in large dark brown bottles that sucked cold sweat from the humid air. I don't remember much about the food we ate—just that it was delicious and the servings were generous. The chilled beer (*uno mas cervesa por favor!*) flowed, and all of us

Opposite: A Black Howler Monkey

31

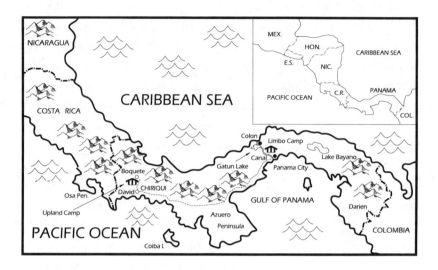

students were happy and, yes, carefree. Hey, we were on a field course paid for by the university and we were headed into the jungle!

To get to our camp on the Pipeline Road, we took the canal train—a small ramshackle thing—to Gamboa, where we were picked up by Neal Smith, a biologist based at the Smithsonian's Tropical Research Institute. Smith is famous for his work on mate selection by gulls in the Arctic and cowbird nest parasitism in Panama. We drove to Limbo Camp, a rustic research site along the Pipeline Road, now a part of Chagres National Park. Limbo Camp was established by James Karr, an ornithologist studying bird populations in the lowland rainforests there. Limbo had become a popular forest hangout for tropical biology freaks and graduate classes.

When we arrived at Limbo, it consisted of nothing more than a corrugated metal roof over a rectangular concrete floor, a clear stream for washing and drinking water, and an array of forest trails. That was all we needed, really. We put up our little mountain tents under tall shade trees and established our encampment in short order. Charlie Munn, another graduate student, and I, aided by several others, began setting up a line of mist-nets in the forest across the road. Given my strong opinions and overbearing (know-it-all) field manner, Munn immediately christened me *Bwana*—a name that stuck

for the length of the trip. I couldn't help it. I loved being in the field, and I
was sure I knew the best way to do just about everything. I was a few years
older than most of the group, and I had just returned from fifteen months in
New Guinea, which colored my world view.

Limbo Camp and the Pipeline Road were a young naturalist's paradise.
I was bursting with questions: "What was that huge tree?" (*Ceiba pentan-
dra*); "What was that loud morning birdsong?" (a species of wood quail);
"What is that bundle of dirty gray fur??" (a Three-toed Sloth) . . . Everything
was new, and so prolific! Vegetation, birds, insects—the place was *bursting*
with life. It is safe to say there is no comparison between the forests of Papua
New Guinea's Upper Watut and those of the Pipeline Road. Pipeline is so
much richer, the life forms so much more developed and exuberantly elabo-
rated. Here was tropical nature as Walt Disney would wish to portray it. By
contrast, the hill forest in Papua New Guinea is much more akin to a minor
Impressionist painting—lovely to behold, but by no means as breathtaking.

Given how much a creature of the daytime I was, I liked to challenge
myself by spending time out in the forest at night. The sensory demands
and the transformation of time and space are powerful. Navigating a trail in
the forest at night without a light gave me a new perspective on the forest
and its denizens. Breaking the stranglehold of fear that tends to overwhelm
a person "lost" in the dark is liberating.

On that first night at Limbo I suggested a walk along the road and
through the forest. A few of the class members joined me. In the dark forest,
once we left the openness cast by the starry heavens, the world closed in
around us. Distances stretched out. Noises magnified. And insects, the true
owners of the planet, took over. Layer upon layer of insect sounds woke the
night; katydids and their kin made a racket that seemed never to end. Frogs
chimed in, dissonantly, in cadence, and the occasional night bird added to
the din. A sudden thrashing in the brush made the hair stand up on the back
of my neck. We turned on lights and flashed. We saw nothing. Whatever
had caused the sound was gone in a hurry. A few minutes later, stopping to
listen to the mournful note of an owl, I looked down to see that I was among
a legion of army ants, whose column was on the move. The energy of that

mass of life was imposing. Some of them swarmed up my leg and I danced about involuntarily, plucking them from my socks and boots. The soldier ants have oversized jaws that give them a fierce look. Although we had hopes for an Ocelot, our walk ended without any mammalian encounters. I slept well after that evening stroll in the jungle!

The next morning, dozing in my little blue-and-yellow tent, I was jolted awake by the staccato, croaking roar of a Black Howler Monkey. The only comparable sound is the repeated cough of a Leopard. The howler is one of the characteristic natural sounds of the Mesoamerican rainforest. Upon hearing it again in 2000—twenty-two years later—I was again electrified. There is nothing I can think of that better epitomizes this environment, except perhaps the ringing, repeated song of the wood quail, which I heard for the first time a minute or so after the first roars of the howler. The televised nature specials that feature tropical forest wonders stress visual pattern and color. They do not capture the remarkable aural tapestry. The night sounds and the dawn sounds, even the midday sounds, are a key part of this environment's richness.

Even the silences were special. There were times, wandering a forest trail, when I was captivated by the peaceful sort of silence that could settle on the forest when the nearby bird flock had slipped away, or when the howlers were napping, and when the cicadas were resting their stridulators. I would hear instead the riffle of water from a nearby stream, or a light breeze rustling the canopy, or the musical wings of a flock of pigeons coursing over the forest.

During my sojourn in New Guinea, I had not experienced the special pleasure of a forest with monkeys in residence. Primates range the forests and woodlands of Africa, Latin America, and Asia, but have been halted in their global expansion by Weber's Line, an ancient deep-water barrier just east of Sulawesi in the central Indonesian archipelago. In Panama, I was fascinated to learn just how primates fit into the rainforest equation, and I wanted to learn what a more typical rainforest was like. Our second morning on the

Pipeline Road, we discovered a tall observation tower in the forest nearby, a 100-foot structure made of aluminum tubing, which rose into the forest canopy. Atop the tower was a flat metal platform where we could sit and take in the splendor of the treetops. The remarkable microenvironment was little known and little studied in 1978. It is a bit like the top side of a cloud bank—highly irregular, diverse in shape and color and form, and in places incomplete. This is the business end of the rainforest, where so much happens: flowering, fruiting, fruit eating and dispersal, pollination, leaf eating, predation, and territorial display by birds and mammals. In 1978, study of the forest canopy was a field just being invented by Meg Lowman, Nalini Nadkarni, and other canopy lovers. Today it is a full-fledged realm of study.

The naturalist, trapped by gravity, spends most time in the forest shadows at ground level. From that lowly vantage point, it is difficult to see much. It is a neck-craning inconvenience to be earthbound when all the monkeys and all the most interesting birds are hanging out up top! What a relief to climb into the bright sunlight and encounter the rich color and texture and action. Butterflies, all sorts of lesser insects, swifts and soaring hawks, flocks of parrots, solitary hummingbirds, and much more all parade in the outer shell of top vegetation.

On my second morning atop the tower, something caught my eye far off to the right—black-and-white forms in the shady subcanopy, moving surreptitiously. Big forms. Certainly not birds. I focused my binoculars on one—a monkey! My first glimpse of a wild primate. Here was a troop of five Common Capuchin Monkeys, known familiarly as organ-grinder monkeys. This species is one of the more attractive of the smaller New World primates, its lovely white face framed in black. The capuchins I was watching moved quickly and quietly through the subcanopy, passing fairly close to the tower, but showing little or no fear.

Human observers have a particular affinity for primates, for an obvious reason; it is a bit like looking at yourself in a mirror. Monkeys, in particular, are like miniature stuffed-toy versions of people, at once comical and endearing (from a distance). Although some primates range into temperate zones north and south of the equator, most of this diverse group inhabits tropical

forests. They are emblematic of the jungle. Many forests support as many as
five or six species, some even more. Their presence enriches any rainforest.
The Neotropics have their tiny marmosets and tamarins, squirrel monkeys,
woolly monkeys, spider monkeys, titi and saki monkeys, howlers, and night
monkeys. Africa has great apes (chimpanzees, bonobos, and gorillas), ba-
boons, mandrills, colobus and guenon monkeys, mangabeys, vervets, and
bushbabies. Asia is home to the diverse macaques (too similar to humans to
be likable), leaf monkeys, gibbons and orangutans, and the tiny, lemur-like
tarsiers and lorises. Madagascar is the sole residence of the plush-toy-like
lemurs, which are the most beautiful of all the primates.

Having wild primates in the forest is rather like adding sprites or lepre-
chauns or elves into the environment. I walked the trails in Panama with a
sense of being watched by the little people. It is something I could not feel
in a forest in the Pacific.

On this field trip I was a bit smug because I was working the forest dressed
for comfort, in flip-flops, bathing suit, and T-shirt. My school comrades were
dressed in boots, socks, long pants, and long-sleeved shirts. Not only were
they hotter and stickier than I was, but they were beginning to show the red
rash of infestation by chiggers—the tiny bushmites that had given me fits in
my camp in the Upper Watut. I myself had no evidence of chigger attack, so
I surmised that my scanty dress served as protection. Chiggers tend to home
in on the dark, damp, tight places confined by boot, belt, or tight clothing.
Why they do so is a bit of a mystery. Perhaps the pressure makes the blood
rise into the capillaries near the skin's surface and thus be more readily available to these tiny red bloodsuckers.

In any case, our Limbo Camp group was suffering badly from the
chigger plague. Some colleagues were absolutely miserable with the itching.
I was itch free and happy as a lark. But, my pleasure was short-lived. After
a couple more days, the scourge visited me in a big way. I awoke in the wee
hours of predawn, my fingernails red with blood from clawing in my sleep
at the itch in my ankles. Alone in my tent, the air temperature tepid but not

The 1977 tropical ecology trip to Panama. Front row (*left to right*): Hannah Suthers, Mary McKitrick, Mary Ellen Curtain, Cathy Bristow, Joan Aron, Bruce Beehler. Back row: Robert Knox, Howard Wildman, the Hartmann brothers, John Terborgh, Robert Dahl, Tony Janetos, and Charlie Munn.

cool, I was on fire with itching. It makes my skin crawl as I write this, thinking back to that hideous night.

What to do? I got up, took some aspirin, and walked the darkened Pipeline Road, seeking breezes that might quench the itch. I hobbled to the nearby stream and doused my legs in the luxury of cool, soothing water. That helped, but it did not solve the problem. I must have spent more than an hour with my legs soaking. While sitting there in the dark, waiting for nothing at all, I heard a scuttling, splashing, rustling a few yards downstream. I quietly pointed my headlight toward the sound and gently pushed the button. There, flooded in the light beam, was a curious little creature, entirely oblivious to my presence or to the light—a Water Possum. It is the only American possum adapted to semiaquatic life, and until that moment I never knew such a creature existed. It looked like a cross between a large rat and a small

New Guinean marsupial, but a pattern of obscure pale stripes crossed the dark pelage of its upper surface. Not surprisingly, I was the only person in my group to encounter this animal. All I can remember now is a smallish, short-haired, incurious creature, moving up the far edge of the small stream, its small forefeet working the stones in search of prey—reminiscent of a raccoon. The encounter eased my pain for a few moments.

All of us itched the rest of the trip. There was no way to undo the damage of the chiggers, once they had done their nasty work.

In the winter of 1958, when I was seven years old, I fell in love with woodpeckers. I began looking for books on woodpeckers at the main branch of the lending library in downtown Baltimore. I did locate a lovely popular book by Fanny Eckstrom (*The Woodpeckers*) and a more technical work by Arthur Cleveland Bent (*Life Histories of North American Woodpeckers*). Of all the woodpeckers, the one I loved most was the Ivory-bill. No big surprise. This extinct or near-extinct creature is an icon for many ornithologists. What's more, in the spring of 1959, on a family trip to tidewater Virginia, I glimpsed a large black bird from the window of our moving car as we left the parking lot of historic Jamestown colony and I quickly surmised I had seen a female Ivory-billed Woodpecker. Knowing the significance of such an observation, I immediately informed the occupants of the car (my mother, my maternal grandmother, my brother) that I had just made a major ornithological sighting. No one doubted my identification, and there was some discussion about how to relay this scientific finding to the Audubon Society.

Whatever I saw that day, you can be certain it was no ivorybill. And I did not press my claim. I subsequently thought I would never in my lifetime see an ivorybill, for I did not know there were other species of ivorybills in the world. So it was with no small pleasure that I glimpsed, from the Panama tower, a large black, white, and red bird—an ivorybill named the Crimson-crested Woodpecker. Not as large or as elegant as the mythical North American ivorybill, this was still a great bird. A pair of them were working over some canopy branches not far off, and one of the birds gave a

tinny, high-pitched series of notes, *kuduk kuduk kuduk*. This male was gar-
ishly plumed with a flame-red head and crest, a black back broken by twin
white slashes on the shoulder, and a profuse black-and-buff barring on the
underside. Later in the morning I heard one of these birds give a distinctive
signal drumming on a dead stump.

Seeing the Crimson-crested Woodpecker made me pine for our American ivorybill, the ghost of our southern swamplands. At the time of our
Panama trip I was convinced that this marvelous bird was long gone from
U.S. soil. I held out hope that a population might be hanging on in the
isolated mountains of the Sierra Maestra of Cuba. Certainly I would never
have predicted its rising from the ashes of our laid-to-waste bottomlands,
most of which had been clear-felled repeatedly or converted to agriculture
or tree plantation. I tip my hat to John Fitzpatrick and his Cornell team for
their persistence in following up the rumors that emanated from that remnant
Arkansan swampland. Although at the time of this writing the number of
Ivorybill doubters is growing, I am not among them. I take heart from the
new report emanating from western Florida. At the very least, this bird is a
tantalizing will-o-the-wisp that will continue to haunt our dreams. I am not
eager to see the Ivorybill declared "gone forever."

A pleasant clear-water river coursed through the forest and approached the
road not far from our camp. We found a nice deep spot where sun and
shade dappled the water. This was our water hole, and when we were not
atop the tower or checking the mist-nets, we visited the water hole. Here we
could soak our festering chigger rash and wash away the grime and sweat.
I would lie back in the water and feel temporarily at peace. Looking up, I
could usually find a swarm of dragonflies hovering above—some sort of mating assemblage? Birds passed by too, and more than once when somebody
shouted the name of some rarely seen bird, one or more of our group would
quickly splash from the water to get towel and binoculars. As I learned in
the Upper Watut, wherever one works in tropical forest, a water hole is a
must. It's what makes work in the forest bearable. Sitting in the water on the

clean sand or gravel, with tiny fish nibbling at one's toes, a motmot or barbet calling in the background, a Blue Morpho flickering from blue to leaf-brown as it lazily moved among the verging vegetation, is a pleasure only to be had in a tropical forest.

Another phenomenon I had not encountered in my New Guinea days was leafcutter ants. I came upon them while walking a forest trail. At first all I could see was small, rounded pieces of green leaf moving in a line from one side of the forest track to the other. On closer inspection it became clear that each bit of leaf was being toted by a single ant, the leaf held erect like a tiny sail. The pieces waved back and forth a bit as the ants earnestly struggled forward with their loads. So many things about leafcutter ants were remarkable. First, that the ant was able to cut a tidy piece out of a living leaf. Second, that it could actually manhandle the leaf, which dwarfed the ant. Perhaps most remarkable of all was that the ant used the leaf to construct an underground fungus garden—the leaf serving as substrate and food for the fungus. The ant is a communal gardener. The fungus that is produced is fed to the ant larvae.

In Limbo Camp, each day had a rhythm of sorts, centered on eating, managing the mist-nets, and making observations in the forest. To a group of undergraduates and graduate students, eating was easily as important as the other activities. We shared cooking and cleanup responsibilities. Rice, beans, and onions were the dominant components of many meals. Nobody complained, because we all were always hungry. Meals and almost everything else were a kind of controlled chaos; it was fun, more of a holiday than a class. But we were continually learning, continually questioning, continually what-iffing. The mix of teacher, grad student, and undergraduate is a remarkable concoction that can produce valuable interchange, lots of ideas, and unconventional learning. It is a great way to break down the barriers that arise on campus. John Terborgh over the years gave a number of students their first glimpse of the tropics through his winter field courses.

One day several of us bumped into a carload of gringos coming down the Pipeline Road (uncommon, because the road was gated at that time). The group stopped to chat, and it turned out that they were from the Panama Audubon Society, out conducting a Christmas bird count. What a place to do that! As we compared notes about the day's bird sightings, one of the gringos motioned to the roadside and pointed out what appeared to be a moving pile of dead leaves. It was, in fact, a Three-toed Sloth. We all sauntered over to have a look at this hapless beast. It moves at glacial speeds, even when under duress, and seems slow-witted as well. It more or less ignored the crowd and crossed the road, ever so slowly. As it did so, we discussed the habits of sloths and one person mentioned that they were typically infested with ecto-parasites as well as algae that grow in their hair. One of the more philistine of the birding group took it upon himself to "help" the sloth by spraying it with insect repellent. It was, in itself, a repellent act, but there was no stopping the fellow. There was something pitiful about the docile creature, and a human's compelling need to interfere with nature. In the same way, people living in rural America like to organize rattlesnake roundups.

During my first stint in Papua New Guinea, instead of studying the birds of paradise to which I had been assigned, I spent most of my time trying to gain an understanding of the environmental specializations of the many species of New Guinean honeyeaters. These are a large group of songbirds, perhaps the most characteristic birds of Australia and New Guinea (although the kookaburra is better known because of its mention in the Australian song *Waltzing Matilda*). Honeyeaters are dominant foragers at trees that produce flowers or small fruits.

Thus, when I was beginning my tower watch in Panama, I was immediately attracted to the aggregations of small birds visiting a large flowering *Luehea seemanii* tree. This phenomenon included a lovely, gemlike group of birds—the neotropical honeycreepers. I thought it might be worth comparing their feeding behavior with that of their honeyeater counterparts in the Pacific. I immediately resolved to make systematic observations of the honeycreepers visiting the flowering *Luehea,* to compare with data I had collected in Papua New Guinea in 1976. The honeycreepers I was observing were compact and colorful—almost analogous to hummingbirds, except that they are larger and do not hover. They are among the most beautiful of Panama's birds, various shades of blue or green, patterned with black, with brightly colored legs.

The birds visiting the *Luehea* were primarily four species: the Green Honeycreeper, the Blue Dacnis, the Shining Honeycreeper, and the Red-legged Honeycreeper. These four I observed visit the flowering tree repeatedly every day, and it soon became obvious that the foragers followed a regular pattern of resource use. The Green Honeycreeper visited as a male-female pair and was territorial, staking out a piece of the canopy and defending it from other species. The other three species came and went as a flock, and seemed to share the flowering resource. Thus, I would see the male and female Green Honeycreepers spend time foraging alone in a single sector of the *Luehea;* from time to time, a group of the blue honeycreepers would descend on the tree and forage together, avoiding the sector defended by the Green. This was, by and large, an orderly process.

What I had seen in New Guinea was strikingly different. The honeyeaters there were pugnacious. The foraging at a flowering tree was dominated by fighting and disorder, with neither mixed-species flocks nor orderly territoriality. In general, the larger species beat up on the medium-sized species, who beat up on the smallest species—a sort of dominance hierarchy mediated by size. The pattern seemed to signal a major difference in nectar-feeding between the Pacific and the Mesoamerican neotropics. Future studies may confirm this as a general pattern. At the time, I was excited to make field observations that were translatable into one of my first published scientific papers.

I guess it is fair to say that science is any study that one can have published in a scientific journal. That may sound cynical, but it highlights a main point of the interaction of science and research—the peer-review process. Although the "mad scientist" portrayed by Hollywood is invariably a loner, science is

rarely carried out in isolation. The whole point is not so much pure achievement, but rather illuminating a problem of interest to one's colleagues. Thus
the real achievement is the selling of the work and its significance to the field.
That is not always so easy.

Becoming a working scientist involves the considerable transformation
from a fresh student recruit to a graduate student. That evolution is one of the
main purposes of graduate school and postdoctoral study—something like
morphing from larva (graduate student) to pupa (postdoctoral researcher)
to imago or adult (research scientist). The student is a sponge soaking up all
sorts of information and trying to make sense of it, and at the same time sort-
ing through subfields and hypotheses in search of a unique line of research.
The initial bits of original work are completed for a Ph.D. (usually less product is generated than the student had hoped). Then at the postdoctoral stage
the fledgling researcher tries to delve into that problem and generate new
thinking and new data that will enlighten the world. Finally, as a professor,
one attempts to create a research team (of undergraduates, graduate students,
and postdocs) that, with the professor's leadership and fund-raising skills,
generates an uninterrupted series of results, a body of work that can change
the world's view of a subject and move science forward. In 1978, I fervently
aspired to be part of this process, but I was uncertain where I would fit in.

The trip to Panama helped me move a step or two forward on this career journey. My discovery of a field phenomenon (the honeycreeper nectar-
feeding) that could be compared with a similar phenomenon in New Guinea
allowed me to think about systems and comparative analysis. My interaction
with a diverse group of young biologists allowed me to test many of the ideas
I had generated in isolation in New Guinea, in a nonthreatening informal
setting. In addition, group discussions allowed me to see more clearly what
was a valuable research topic and what was not.

A second-year graduate student tends to focus on "favorite creatures"
or "favored places," rather than on cutting-edge questions and hypothesis
testing. The student practices doing science by first learning how to collect
data. Only later, with intellectual maturity, is it possible for the student to sort
among the random noise of nature to find a trend that might shed light on

a general question. Highlighting the trend may result in a mildly interesting paper for a niche journal. Answering a general question may ensure a series of key papers that lead to tenure at a major research university.

Universally, we second-year graduate students at Princeton had great difficulty identifying a tidy research project that would serve as a dissertation study. Some of us were competent naturalists and knew a lot about birds or plants. Others were reasonable theoreticians or lab types. But few of us knew enough of both nature and theory to identify a cutting-edge biological issue and match it with a natural phenomenon exhibited by some plant or animal in nature, in a way to get the issue clarified in the two or three years available to a student.

I suspect that identifying a sexy but tractable problem and matching it to solid field data remains the hardest part of earning a graduate degree in biology. I still hear from budding grad students who want, instead, to "learn everything" about a particular charismatic species (like a howler monkey), or perhaps document all the species in a particular patch of rainforest. That is natural history, not science.

After a week at Limbo Camp along the Pipeline Road, we headed west to Panama's Chiriquí Province, adjacent to the border with Costa Rica. This move required a bus ride to the city of David, followed by a drive up into the mountains to the small town of Nueva Suisse. Here we settled into a field camp in a patch of upland forest near Boquéte at about 4,000 feet above sea level, verging on a shade-coffee plantation owned by Radibor Hartmann. It was cool and relatively dry, and apparently free of chiggers.

The upland forests of Chiriquí offered another splendid new environment for a wide-eyed graduate student. The natural forest had been much reduced and badly hacked up, but significant patches remained. Birds and mammals were still diverse and common. It was a world apart from the Pipeline Road. The forest-edge and coffee plantings abounded with North American migrant warblers—Golden-winged, Tennessee, Cape May, Black-and-White . . . wonderful birding with little effort. The marvelous mix of

This Emerald Toucanet, netted in the forest, was a
vicious combatant when being removed from the net.

North American wintering birds and local Mesoamerican resident species
presented lovely new combinations of song and color and behavior.

Señor Hartmann had told us that a Jaguar still prowled the forests
around his estate, and that with some diligence we might observe this species
at night. I was captivated by the idea of seeing a Jaguar, though I recognized
that it was a long shot. I hatched a scheme to increase my chances.

I decided to stake out a water hole, as it was the dry season. I located
a clear, rocky pool in a nearby small stream. The stream flow was low, so the
best place for game to drink was at the pool. I picked a medium-sized tree
with decent branching to serve as my observation stand. At dusk, I would
climb the tree, tie myself onto a branch, and silently wait in the dark, my
headlight at the ready. This is not something I had done before, but I was
hopeful that it was the right way to catch a glimpse of the carnivore king of
the Panamanian forest.

After an early supper I made my way to the water hole. As I started to
climb the tree, I found that getting up to a decent branch was more difficult
than I had anticipated. Using a loop of rope, I managed to hitch my way
up. Then it was a matter of making myself minimally comfortable and tying
myself in so that I would not tumble if I fell asleep.

The moon was nearly new, so the night was very dark. The air cooled quickly and the chill set in. At 4,000 feet it gets quite cold on a cloudless night. I was seated crossways on the large limb, my back propped against the tree trunk. For some reason, I had brought neither a sweatshirt nor a pillow to sit on. Within ten minutes I was uncomfortable. Within twenty minutes my back and my bottom were in pain from the hard and irregular surfaces of the tree. I kept adjusting, trying to make as little noise as possible. I could not get comfortable for more than a minute at a time.

Then the insects came. Crawling insects. Ants, hoards of tiny ants, and other creepy-crawlies sought me out. I suppose I was a novelty, and I probably looked or smelled like food! Ants and other insects crawled all over my clothing, searching for passages into my warm undergarments and bare skin. It felt as if a whole ant nest had colonized me. One saving grace was that they were not biting, but they certainly annoyed and tickled. A large bug made its way onto my back. When I reflexively slapped at it, it turned out to be a stinkbug.

The dark and the silence penetrated my consciousness, creating a strange mix of nervousness and boredom. All sorts of unidentifiable sounds wafted in from the surrounding vegetation. I periodically checked my watch. Ten minutes seemed like a half hour. After an hour or so, I started to drift off, in spite of the ants and discomfort. It was, after all, night, and I had spent the whole day walking and observing. I drifted in and out of something like a state of delirium, with weird dreams rising and falling in my consciousness.

At ten-fifteen I was startled by a loud sound in the undergrowth. Some sort of mammal was definitely below me. I could hear leaves being scrunched at intervals. I tried to pre-focus my headlamp in the direction of the sound, but when I turned it on I was blinded by the sudden flash. The thin green understory reflected back all of the bright light, and I struggled to pick out any creature that might be skulking in the undergrowth. Nothing.

I fell asleep and woke again to some sound in the night. I breathlessly repeated the exercise. I flashed the flight on and . . . again could see nothing.

My body was cramping up and I was starting to suffer hypothermia. I gritted my teeth and held on. Time passed. Then there was a loud rustling downstream. Again I switched on the light, which precipitated additional sounds of a quadruped retreating into the night. Not seen.

My backside was incredibly painful. I have never spent so much time outdoors at night with so little profitable return. I heard nothing. I saw nothing. After four interminable hours I called it quits and dragged my cramped body back to my little tent, which by comparison was a paradise of comfort.

I was given considerable (much-deserved) razzing by the group. What a stupid stunt, with minimal chance of success. But . . . but . . . I wanted badly to see a Jaguar.

In fact, mammals were fairly common in the forests of Chiriquí. I saw my first Red Spider Monkeys in the high canopy on the day of our arrival; it was a glimpse of one of the most acrobatic primates of the New World—but only a glimpse. Working in a forest with these wonderful creatures adds a certain something. Just knowing they are sharing the forest with us humans has an impact—an "existence value," even though it may be hard to put a price on.

On the last day of our stay, I came upon a remarkable weasel relative in the branches of a *Cecropia* tree beside the trail I was walking. It was a Tayra. I suppose it could be considered a tropical, short-haired version of our Wolverine or perhaps an oversized Pine Marten. The Tayra was darkly furred with some frosting around the throat, and about 3 feet long including its long bushy tail. I stopped and watched in fascination as the animal searched the tree in vain for prey, then descended to the ground, stood on the trail, looked at me impassively, and slowly moved off down the trail seemingly unconcerned with my presence. As with the water possum, I had not known of the Tayra's existence until I observed it first hand. In 1978 there was no handbook of the mammals of Panama to study.

The most remarkable primate I encountered on this trip was our professor, John Terborgh. A world-renowned tropical ecologist, Terborgh is perhaps most famous for his establishment of the Cocha Cashu biological station in

Manú National Park in Peru—a fertile breeding ground for a generation of cutting-edge tropical biologists. Also one of the founding fathers of conservation biology, Terborgh has published influential work on community ecology, behavioral ecology, conservation biology, and management of protected areas. On this field trip, he was certainly the dominant male of this temporary social assemblage. We all spent a lot of time listening to John talk about what we were seeing, and of course we were continually asking his opinion on all sorts of issues in tropical ecology. Here John was in his element. He is first and foremost a man of the neotropical forest.

John Terborgh is a person of many talents and always entertaining to be with. He has a passion for nature in all its forms, and strong opinions about everything (as do virtually all of my male tropical ecology colleagues). His leadership of the field class was remarkably free; he let his grad students sweat the details and kept his own eye on the big picture. I never saw him lose his cool, and he never meddled in the details of the day-to-day management of the details of survival. He knew, with this group of can-do students, that there would be no problem ensuring the daily duties were carried out to everyone's satisfaction. Students practically fought over the right to make dinner. The sweaty and onerous task of cutting trails for net lines, followed by erecting the mist-nets, was done without complaint. John had the devoted energies of the whole class, and it was something he had come to depend on. During his tenure at Princeton he led one of these classes annually, and he had grown accustomed to camaraderie and faithful effort on behalf of the course.

Another singular individual, in many ways like John Terborgh, was Charlie Munn. He too was one of Terborgh's students, studying the behavior of Amazonian mixed-species bird flocks. I felt a natural affinity for Charlie for several reasons. He was a very likable fellow. He loved studying rainforest birds. And he grew up in Baltimore, where we attended the same grade school. It was Charlie who christened me *Bwana* in his irreverent but friendly way. He was my chum on this trip, and we spent a lot of time together comparing notes about fieldwork, ideas, and the future.

After graduate school, Charlie went on to become a research scientist with the Wildlife Conservation Society (WCS), where I joined him as a fellow researcher for several years. At WCS, Charlie became a crusading conservationist in search of workable solutions to the thorniest issues in tropical conservation. He has since branched out on his own. In pursuit of neotropical conservation solutions, he has founded several local organizations that work to conserve tropical forests through innovative and self-sustaining initiatives. There is no question in my mind that the passionate work of dedicated and driven people like Charlie will make all the difference in these last wild places we all cherish.

On this visit to Chiriquí I learned firsthand about what is today known as shade coffee. Radibor Hartmann's plantation was mainly coffee planted under a natural canopy of original upland forest. The forest understory had been thinned, with the largest canopy trees left to cast shade on the coffee plants grown beneath. In the early 1970s, John Terborgh and John Weske had noted that plantations of shade coffee supported rich assemblages of wintering neotropical migrant songbirds that nest in the forests of the northern United States and Canada. Thus this shade coffee is great for birding, and, more important, is great secondary habitat for "our" favorite songbirds during their long winter "vacations" in the far South. With the massive loss of original forest, the shade coffee serves as a substitute home for these birds— and is apparently critically important for their survival.

After about a week in the lovely shade coffee land of Chiriquí, we were back in Panama City to catch our flight home to the United States. Our short layover in the big city was fun—dinners in seedy restaurants with great cheap food and icy beer in those tall, hefty, returnable bottles. We celebrated the successful completion of our course, reminisced about what we had seen, and really did not look forward to our return to winter in Princeton, where the snow was piling high. We boarded our plane in Panama dressed in shorts, T-shirts, and sandals.

One of the striking aspects of Panama City in 1978, even to a short-term

visitor, was the military presence. Every downtown street corner was deco-
rated with a soldier in green fatigues and an automatic weapon. This decor
apparently was the "management style" in the days before George H. W.
Bush's war against Manuel Noriega. We paid it no particular mind, but noted
that it was there, nonetheless. The canal still lay in a U.S.-run zone that was
soon to be given back to Panama through the execution of a treaty signed by
President Jimmy Carter in 1979. In 1978 I am sure the canal issue was a sore
subject with the people of Panama, and perhaps the military show on the
streets was a way of putting pressure on the United States for change.

In 1997, I was working for the U.S. Department of State and was assigned to
watershed protection issues surrounding the Panama Canal. Undersecretary
Tim Wirth was pushing the importance of environmental security issues, and
the canal watershed was one of them. Finally—a high U.S. government official
who understood the ties linking environment, forests, and global security.

Here was the issue in brief: the canal was a strategic transportation
route for the United States. If the tropical forests of Chagres National Park
were cleared, then the water needed to raise the canal locks daily would
be in short supply during Panama's dry season. No water, no canal traffic. Conserving the watershed forests, then, was critical to "conserving" the
Panama Canal. Deforestation could lead to serious economic and military
breakdowns that could have global implications.

Here was a prime example of how the environmental services provided
by a tropical humid forest underlay a key enterprise for a developing nation
and the world at large. Tim Wirth's focus forced Panama to take this issue
seriously and taught a lot of decisionmakers about the important linkage
between the environment and enterprise.

I was briefly in Panama again in May 2000, on my way from Nicaragua
to Jamaica (the long way around). Panama in 2000 seemed another world
from the one I had seen in 1978. The airport was large and very modern. I
could send and receive email there. The military was not in evidence. Pros-
perity was everywhere. It was a new Panama.

❧❧

Tropical humid forests are found the world round, but their intercontinental differences are as significant as their similarities. On the one hand, jungles look alike from New Guinea to Panama: they tend to be dominated by a high canopy of hardwoods, an abundance of vines and lianes, and a mixed understory that features saplings, palms, and tree ferns. Their soils are typically thin, and their plant composition is usually patchy at the local scale, the product of various sorts of natural disturbances over time. But that is where the similarities end. A few widespread tree families and genera feature prominently in rainforests around the world, but forests from one continent to the next tend to have very different trees at the genus level and below.

New Guinea and Panama are about as different as any two tropical humid environments on earth. Panama's jungle differs from that of New Guinea because of the presence of army ants and their attendant flocks of ant-following birds; its dominant avian seed dispersers, which include the toucans, barbets, manakins, and cotingas; the infestation of the forest with all sorts of seed predators like monkeys and squirrels and deer; cats and canines as the top carnivores.

In New Guinea, birds dominate the forest rather than mammals; seed dispersal is carried out by soft-gutted pigeons, birds of paradise, bowerbirds, hornbills, and a large array of fruit-bats, including the giant flying foxes absent from Panama; carnivores and top predators are essentially absent; the top seed predators are cockatoos and rodents; there are no deer, squirrels, cats, canines, or monkeys.

The presence of monkeys in the Panamanian forest is perhaps the feature that most distinguishes Panama from New Guinea. Monkeys, because of their physical and mental abilities, change the ecological equation, especially with regard to forest regeneration. Because primates can use their hands and teeth in so many ways, they serve as pervasive seed predators that strongly influence plant strategies for flowering, fruiting, and seed production. Monkeys make a mess of any delicately evolved fruit (or flower, for that matter), so plants need to adapt to this impact in Panama. Squirrels only add to the problem. Each seed is a plant's hope for the future. Monkeys and squirrels, in most instances, look at fruits as packages that include edible seeds. If a plant cannot have its seeds dispersed and germinated in the forest, it will not endure. The presence of primates in a forest changes the equation in other ways. The specialized coadaptation of food plant and seed-dispersing birds of paradise that I have studied in New Guinea could never happen in Panama because of the interference of monkeys (and squirrels). These are two distinct worlds.

Another major difference between New Guinea and Panama is cultural. New Guinea remains a world dominated by indigenous and traditional peoples and forest-dwelling societies. By contrast, Panama is a society dominated by Hispanic colonial culture and practice, with indigenous and traditional peoples playing a subordinate role. This factor has greatly influenced economic development and especially land tenure. New Guinea has millions of forest-dwelling families living on their own traditional lands. Panama has evolved as a Western society, with few indigenous peoples living on their ancestral lands. The landscape has been transformed because of Western agricultural practices and the near-universal development of cash-cropping and plantations. Both occur in New Guinea, but they are localized and concentrated near urban and coastal areas.

One of the ironies of shade coffee relates to the U.S. government and its attempt to help the poor of Latin America. For a number of years the U.S. Agency for International Development (USAID) funded programs that sought to replace the "old-fashioned" shade coffee with high-tech "sun" coffee to boost yields and thereby reduce poverty in the developing countries of Mesoamerica. This policy resulted in the clearing of thousands of acres of upland forest and the creation of monocultures of sun coffee—whose productivity was maintained by expensive fertilizers and herbicides. Only in the early 1990s did USAID have a change of heart, funding a number of shade coffee projects because of their environmental and community benefits. So many do-good projects funded by government development agencies around the world end up doing harm. It is comforting to know that USAID was able to make the change in coffee management before it was too late.

What we have learned in the 1990s with regard to coffee growing is that "old-fashioned" is not inferior. The conversion of much of the world's coffee and tea estates to shade-free "sun" plantations was unfortunately "modern" technology that was supported by all sorts of donor governments. It was later found to be wanting because of the serious impact on local communities, upland watersheds, and wildlife. Radibor Hartmann had it right in 1977.

CHAPTER THREE

On the Trail of Ripley and Ali

I recall the thrill of seeing the forest roads covered
with the pug marks of tiger, panther, bear, and
other wildlife each early morning as I tramped along
them behind the local Chenchu tracker.

Sálim Ali,

The Fall of a Sparrow (1986)

BY LATE 1981 I HAD TAKEN A JOB at the Smithsonian Institution, working as S. Dillon Ripley's scientific assistant, based in his research laboratory in the Museum of Natural History. Ripley was an ornithologist famed for his work on the birds of India and the Far East. For twenty years he ran the Smithsonian, making it one of the world's great research institutions.

I initially linked up with Ripley by chance. In late 1975 I was at the Wau Ecology Institute in the hills of central Papua New Guinea. One evening I was working late in the office. Looking through the drawers of the desk for typing paper, I happened upon a letter from Ripley to a colleague of mine who had been based at Wau. In the letter Ripley made clear he was in search of an ornithological field correspondent based in New Guinea. I had no idea whether my colleague (at that point gone from New Guinea) had ever answered the letter, but I immediately dropped what I was doing and wrote a letter telling Ripley that I would be delighted to be his contact in New Guinea. By return post, I was engaged as a researcher on behalf of the National Museum of Natural History. Subsequently, I served as a summer research fellow in Ripley's laboratory in 1977, and Ripley served on my doctoral committee at Princeton. I have maintained a relationship with the Smithsonian and the Ripley family ever since.

Periodically through the 1980s I was asked by Dr. Ripley to visit southern India to help set up field camps from which we could conduct bird surveys in the wildlife-rich remnant forests there. For these expeditions, Ripley would be accompanied by his entomologist wife, Mary, and in one instance would join his longtime scientific collaborator, Dr. Sálim Ali of the Bombay Natural History Society. This threesome had been making field trips to remote corners of the Indian subcontinent for decades.

The goal of the 1983 trip was to facilitate a natural history survey of birds in the Visakhapatnam ghats of Andhra Pradesh. (In this context, "ghats" are two ancient mountain chains in peninsular India, one in the east and one in the west.) The objective of this and subsequent trips was fundamental: to document the birds that inhabit these little-studied mountain forests and advise the state government on conservation of critical habitat.

Southern India and New Guinea are unlike in many ways. New Guinea has trackless jungles and tiny hamlets; India has burgeoning human populations

Opposite: Tribal women carrying grain in South India

and remnant seasonal forests. New Guinea has a young and fragile government, whereas India has ancient societies and a giant rough-and-tumble
democracy. New Guinea's forests come alive with its endemic bird fauna;
India's woodlands, by their spectacular mammalian megafauna. New Guinea
is dominated by its forest peoples and traditional land tenure systems, and
India by government-dominated lands and the huge and powerful Indian
Forest Service.

My fieldwork in India in the 1980s would give me a new perspective
on fieldwork, bureaucracy, science, and culture. It would round out my field
experience, making it more tropical and less Melanesian.

I touched down in New Delhi at about one in the morning on September 15,
1983. I quickly passed through immigration and customs and was suddenly
out in the thick, scent-laden night air, a fragrant advantage of postmidnight
arrival in a new city and new country.

I was, of course, nervous about coming to India that first time, but on
this night all the signs were benign. The driver who was there to meet me
put me in the back of a boxy, diminutive four-door Ambassador car (which
looked like a miniature version of my great-aunt Ola's ancient 1950s Buick).
We were off to downtown New Delhi.

I was bombarded by the rich night air as it rushed into the open window, cool at this late hour. I picked out the sweet smell of burning cow
dung, then a scent reminiscent of dried tea. The smell of wood fires, and
the attendant low haze, showed their omnipresence in a New Delhi night.
A few stars were visible. The wide avenues were empty except for the half
dozen white cattle that shambled across a broad intersection. As we entered
the city, I saw many dormant yellow-and-black taxis at the roadside as well
as the three-wheeled motor rickshaws. I say dormant because each vehicle
was home for its driver, asleep inside. Within twenty minutes of driving at
breakneck speed, I was at the Hotel Imperial in downtown New Delhi.

Morning came quickly and the combination of bright sunshine and the
musical chatter of Rose-ringed Parakeets woke me. The birdlife was noisy.

A House Crow was feeding a baby koel (a parasitic cuckoo) in a tree by the window. A turtle dove and a small group of *Turdoides* babblers were playing at the edge of the lawn of the hotel's back gardens. Before long I was on a shaded garden patio among the birdlife, taking an English breakfast. Looking up I saw White-backed Vultures, Scavenger Vultures, and Black Kites wheeling overhead, high in the hazy blue-white sky. The Hotel Imperial was a "colonial-style" tourist hotel on Janpath, a famous tourist boulevard. Francine Berkowitz, the India expert at the Smithsonian who arranged my trip, assigned me here because she knew it would "soften the blow" of coming to the subcontinent. Indeed. I felt I was in the lap of luxury. There was nothing remotely like this in Papua New Guinea.

Later in the day I made my way to the fortress-like U.S. embassy to present my credentials and to pick up my sack of "PL-480" Indian rupees. U.S. Public Law 480 resulted in the sale of grain to India with an agreement to repay the United States in local Indian rupee currency. In 1983 the rupee was nonconvertible and nontradable, so the U.S. government had piles of rupees, earning interest, in Indian banks. Because the money could only be spent in India, and since there were few American government expenses in India, this fund was seldom tapped and seemingly inexhaustible. We called it funny money. My Smithsonian field trip in India was entirely supported by these rupee funds.

Toward the end of the day, with Delhi baking in the sun, I made my way back to the hotel and visited the pool for a cool swim. I was distracted by the abundant garden birds: House Crows, Red-vented Bulbuls, House Swifts, Indian Mynas. From a small bush near the hotel, a wren-like Common Tailorbird, its tail cocked, sang its staccato, sewing-machine-like trill. A giant fig tree bearing abundant tiny red fruit drew in two bulbul species and a handsome little barbet known as the Coppersmith (because of its monotonous *tok tok tok tok tok* vocalization). A little striped palm squirrel, attracted by the *pishing* sound I made, came down the branches of a small tree to spy on me. Across the street a dun-brown monkey (a Rhesus Macaque) searched the gutter for scraps. I was struck by the incredible profusion of animal, plant, and human life in this city, all living in peaceful coexistence. It was a far cry from New Guinea, where birds played hard to get and mammals were nowhere to be seen. In India birds are everywhere, cheek by jowl with people. In this Hindu nation, birds are left in peace by a large proportion of the population, whereas in New Guinea the birds (and all wildlife) have been hunted and consumed by people for thousands of years.

I had heard many stories about "Delhi belly" and the gastrointestinal complaints of first-time visitors to India, so I was skittish about the food. But I quickly came to my senses. I feasted that night on mulligatawny soup and a spicy shrimp dish with lentils and rice, washed down with a bottle of Kingfisher beer—heavy, dark, tasty. Beer in big bottles was one thing India had in common with Panama.

The next day, Saturday, was a free day for me, so I did what any sensible naturalist in a tropical country would do—I visited the zoo. I was not there to see the caged animals, but to go naturizing. I was not disappointed. Pond Herons, Red-wattled Lapwings, a pair of Black Drongos, and a White-breasted Kingfisher worked the little pond and its edges. A lagoon right on the zoo grounds was home to a colony of Painted Storks, egrets, and night herons. A Pied Crested Cuckoo flopped atop a small bean tree, sunning itself. I encountered a Lesser Golden-backed Woodpecker, which is like a small and colorful Pileated Woodpecker, with a tinny trilled call reminiscent of that North American species. India was indeed the Bird Continent. And this was truly a zoological garden in the best sense. It was vast and diverse, with dwarf woodlands, a variety of Mughal ruins, exhibits, fetid swamps and streams, trash, clots of overdressed strolling family groups. Everything in excess, including a captivating group of Indian Hoopoes playfully foraging under some scrubby acacia trees in one of the pens, in the blazing September heat of late morning. Sun-baked September, by the way, is *not* the proper time to visit India. I'm not sure how this expedition came to be scheduled for September and October, the cyclone season.

After the zoo, I visited the Red Fort and Old Delhi (the ancient city). Given the many unheralded ruins that I had encountered on the grounds of the zoological garden, I was only mildly impressed by the fort. The obligatory snake charmers and other hucksters were overwhelming. The narrow

streets of Old Delhi were choked with people, refuse, smoky haze, and confusion—this was India in its essence, everything coated with a layer of dust
and grime.

❧

After several days of preparatory chores in the capital city, I arranged for S.
Subramaniam, an embassy official, to assist me in getting five huge cardboard
boxes of field supplies to my destination in Visakhapatnam, in southeastern
India. Things did not go as expected. After putting down in the lush and
golden-green rice-paddied lands of Calcutta airport (known as Dum Dum),
we found that Subramaniam and the boxes had to be offloaded there, to await
the next flight. I flew on to Bubhaneswar (Orissa) and thence to Visakhapatnam (affectionately known as Vizag for short). Vizag was the principal coastal
industrial town in Andhra Pradesh, one of India's little-visited southeastern
states. Typical of India, Vizag at the time was a relatively small urban center
of a million or so (but really more like a town than a city). I was there to meet
my host and sponsor, K.S.R. Krishna Raju.

An ornithological correspondent of Dillon Ripley's, Krishna Raju had
convinced Ripley of the importance of surveying the avifauna of the Eastern Ghats of India—the ancient low range that parallels the eastern coast of
southern India, from Orissa down to Tamil Nadu. The Eastern Ghats are,
in some ways, poor cousin to the much higher, wetter, richer, and better-
known Western Ghats. In the late 1970s Trevor Price had conducted some
tantalizing bird research in the Eastern Ghats (with the assistance of Krishna
Raju) and it was a desire to follow up on Price's work that led Raju and me
to Andhra Pradesh.

Vizag was set on the Bay of Bengal—a quite beautiful coastal setting,
but clearly heavily impacted by overdevelopment. The city was situated in
a natural harbor with shapely grassy hills all around. In 1983, Vizag was
probably one of the world's largest small towns. There was no city center
to speak of, just a vast collection of huts, low offices, drab apartment buildings, and palm-thatched dwellings. The urban poor were wretched, near naked,
living in huts not unlike what one might find in some of the poorer hamlets

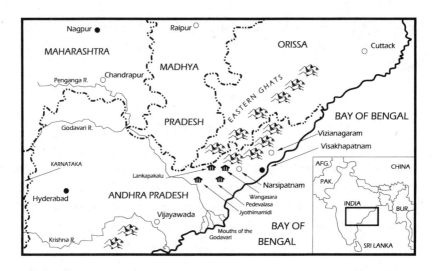

of New Guinea. I was surprised by the primitive nature of it all. The narrow macadamized road from the airport to town was choked with rickshaws, scooter-rickshaws, two-wheeled bullock carts, water buffalo, loose cattle, huge old-fashioned trucks, bicyclists laden with cargo, people tandem on scooters, and the ever-present small Ambassador sedans. Tall billboards touting garish Hindi movies abounded; thousands of cramped little shops crowded the edge of the road.

Vizag boasts one of the finest harbors in eastern peninsular India. At the time of my visit, it was home to India's submarine fleet. Nothing unusual there, except that at the time India was much closer to the Soviet Union than to the United States, and the submarine fleet was supported by the Soviets. There was a certain amount of secrecy about what went on in Vizag; I was told little, although I did see quite a few Russians. At the time, before glasnost and the fall of the Berlin Wall, encountering Russians was quite special.

Back in Vizag after our fieldtrip, I attended an outdoor celebration of the Hindu festival *Diwali*. Many Russians were present, although none made eye contact with me, nor did any attempt to strike up a conversation. I was most struck by the young Russian women, readily identifiable by their striking platinum-blonde hairstyles and abundant makeup reminiscent of the 1930s Jean Harlow. Today, of course, Russians are everywhere and are ever so stylish, but back in India in the early '80s, I was amazed to see these glamorous Russian women at close range, very exotic to an American and seeming from a different era. The Cold War promulgated the existence of two parallel worlds, separated by politics. I suspect the Russians were surprised to see an American in Vizag. I may have been the only one there at the time, because away from the main tourist routes in India Americans were few and far between.

A reserved and polite Krishna Raju met me at the tiny airport and with several Indian Forest Service officials drove me to my lodging, the so-called Ocean View Guest House. It was a horror after the luxury of the Hotel Imperial. Everything was poured concrete. No hangers for clothing, no hot water, no screens in the windows, no view of the sea. The nights spent there were the low point of my trip, marked by mosquitoes hovering over me in the dark, an unsleepable bed with nothing but a thin sheet to protect me.

During my second night at the Ocean View, I was wakened long before dawn by a chorus that included the omnipresent urban pariah dogs and the high chirping of the little palm squirrels. At this point I had spent more than a week in India engaged in preparatory activities. I was aching to get into the forest.

Krishna Raju was a young and handsome South Indian of princely lineage, with a business of his own and a fanatical interest in natural history, especially of the birds of the Eastern Ghats. He had long dreamed of attracting the two scions of Indian ornithology—Sálim Ali, of the Bombay Natural History Society, and S. Dillon Ripley, of the Smithsonian Institution—to his beloved Eastern Ghats. This trip had been planned to accomplish precisely this daunting task. Indeed, Ripley and Ali were due to arrive shortly. I was there to set up a field camp for the two Great Men and be sure everything was ready when they arrived. Krishna Raju and I were to situate and establish the initial field camp; other study sites would be added later.

I visited Raju's home in Waltair, the fashionable section of Vizag. It was

a spacious house set in abundant gardens tucked far from the madness of
the city itself. I met his lovely wife and three pretty little girls, then in the late
afternoon he and I visited the aging Waltair Club, a musty relic of the British Raj. Here was a remnant of the hated British, reoccupied by the Indian
elite of Vizag. It looked to me as if not a stick of furniture had been replaced
since the hand-over in 1947. Quaint does not properly describe the edifice
in 1983. I was beginning to get a feel for the paradoxes and contradictions of
this vast subcontinental civilization.

  In caste-conscious India, the worship of great personages exceeds that
in most Western countries, with the exception of Britain's mania for its royals.
Thus, when word was out that Ali and Ripley were to visit the Eastern Ghats,
the bureaucrats came out of their dark dens, dressed in their finest uniforms,
ready for the kind of photo opportunity that came ever so seldom in central
Andhra Pradesh. It fell to my lot to spend considerable time with a flock of
these government officials over the weeks to come. Some were smart and fun,
others dull and officious. Some knew a lot about wildlife, most knew little or
nothing. They taught me that in south India every workday was punctuated
repeatedly by "taking tea" and "taking food." This world moved at a slow,
steady, predictable pace. And the execution of our field trip was in the vice
grip of the bureaucrats, without whose permission nothing of this sort could
take place in such a xenophobic corner of the tropical world.

I was chafing to get to the exotic-sounding Eastern Ghats, especially because I was stuck in the dreadful Ocean View Guest House. Fortunately, the
next morning, using two official government-issue jeeps, we launched off in
search of forest . . . but first we needed to visit the government headquarters
that oversees the forests and wildlife of this sector of the Eastern Ghats—in
Narsipatnam. Getting to Narsipatnam was an adventure in itself. Whereas in
the United States one zips along on highways, in Andhra Pradesh every road
seemed to be a minor thoroughfare linking crowded neighborhoods. After
wending our way through narrow passages to get out of Vizag, we jumped
from village to village. In the paddy-filled countryside we moved fairly rap-

idly, but we continually dodged around slow vehicles, with the horn tooting almost continually. Indian drivers apply the horn whenever they are near another car, or a bicycle, or a cow, or a herd of goats, or a flock of brightly uniformed school children.

Each tiny town was much like the previous one, clogged with traffic of all sorts (foot, cart, rickshaw), pressed upon one another by the narrowness of the street and the proximity of the shops, all of which seemed to be selling the same goods. Life here was largely an outdoor affair (it was warm the year round, of course). People often slept on rope-net beds (*charpoys*) placed outside the front door or on the street. Toilets seemed to be superfluous; people relieved themselves in convenient places, not necessarily out of sight. Little children played naked in the streets. Young girls made patties out of cow dung and slapped them onto outer house walls to dry (once dry, they were used in lieu of firewood). People bathed in a waterhole or tank (a dammed stream with earthen levies forming a pool) along with water buffalo. Everywhere one looked, there was human activity—people living out the personal details of their lives for all to see.

The stopover in Narsipatnam was the usual drill: office visit, tea break, attempt at pleasantries, strained silences, curiosity of the officials of my family life—was I married? how many children did I have? and the like. Eventually, we were back in the jeeps and on our way into the interior. The ghats from a distance looked like a huge dark wall that rose abruptly from the pale green plain. They apparently were an uplifted plateau with a steep escarpment forming the ghat face that led down to the plain. Before long we were switchbacking up this escarpment, climbing rapidly.

We soon left the open desolation of the plain for a forested environment, which was elixir to me. We drove up, up, up onto the plateau top, and there entered another world, far from the confusion of Narsipatnam and Vizag. Far from the heat and smoke and crush of humanity. *This* was what I had come to India for, I realized. We traveled to Lammasinghi, then Runnagar, then climbed the pass at Sapparla, over to Upper Sileru, a reservoir on the Sileru River. It was the rainy season, so everything was wet and green. The plateau was a verdant land of coffee and pepper plantations, teak, vegetable gardens, and lush remnant forests. It was tribal land, with brightly garbed tribals (as they are termed) walking the narrow roads, heading to market. Their hamlets were tucked in the back country. We stayed at the Electricity Commission's hostel beside the reservoir, a considerable improvement over the Ocean View—and with better food. The remnant forests of terminalia, incense, fig, mango, and bamboo evoked those of New Guinea. I felt at home here, even though the colorful birdlife was Asian, not Pacific. The birds were out and about in abundance: long-tailed treepies, mynas, shamas, rock thrushes, drongos, bulbuls, minivets, wood-shrikes, and babblers.

Dinner that night was a spicy affair, taken on a low table with no silverware. The food was consumed by hand, with the assistance of flat bread (chappati) and rice, pickle, chutneys. To my palate, everything tasted curry-like. Important side dishes, which were to appear at virtually all meals, were the lentil-soupy dal, curds (yogurt), and "salad" (cucumbers diced in a tangy vinegar and chili-seasoned marinade). It was a feast for the senses. The only aspect of local Indian food to which I have never been able to adjust is the bones. Every meat dish (chicken, goat, whatever) comes with the bones in.

Our second day of pereginations brought us to a tiny hill-station hamlet at 2,800 feet elevation named Wangasara, 9 miles by road from Lammasinghi —the site of the ground-breaking fieldwork of Trevor Price. At Wangasara a lovely little forest guest house, property of the Forest Department, would be my home for the next three weeks. It nestled at the base of a forested hill, adjacent to plantations of coffee, pepper, and teak, as well as a mix of secondary forest. I was provided with a field assistant, Mr. V. Ramlingam, who had worked with Trevor Price and knew the ways of bird fieldwork. Although mainly a Telegu speaker, he knew a little English. The forestry people and Krishna Raju had to return to the city, so they bade us farewell and left me and Ramlingam to our own devices. (Remember, their interest focused on Ali and Ripley, not Beehler.) They would bring Ripley and Ali to us, once the Great Men arrived in the province. We had already received a telegram from Ripley indicating some delay in his plans. We had also heard that Ali was sending his great-nephew Shahid to do advance work in preparation for their arrival.

Wangasara was perfect, exactly right for a young field naturalist. I had a field assistant who knew his way around (Ramlingam), a housekeeper/cook (Appa Rao), a house maid (name now forgotten), and an expansive planta-tion of shade coffee and pepper that was literally dripping with south Indian birdlife and Himalayan migrants. I was here in the rainy season, which meant that birds were active and fruiting plants were productive. I could explore the network of trails through the plantation and around the small hamlet, and I did not need to do any real ground-breaking, as the birds were acclimatized to the community. It was like a bird-friendly New Guinea.

It seems that everywhere you go in India has a forest guest house or government hostel of some sort, demonstrating the reach of the Indian civil service (especially that of the Forest Service, a bureaucracy created by the British and continued, to good effect, by post-independence India). The organization was much like a military service, with its considerable pride and dedication, and nifty uniforms too. The management of India's vast forest lands was in the hands of the Forest Service. All things considered, those forest lands were in fair condition, given that the mandate was to make them produce for the benefit of the people of India.

I was thankful for my own tidy forest guest house, which was cozy and well serviced. From dawn to dusk I surveyed birds, and my team single-mindedly supported my fieldwork.

I formulated a plan for a research project: I would investigate how local birds used the different human-dominated environments here in the Eastern Ghats. Right off the front porch was a shade coffee plantation interplanted with Silveroak (an Australian exotic in the family Proteaceae), upon which pepper vines were planted. On the hill behind was regenerating natural forest that had been logged selectively. Down the road was a monoculture of teak. All I needed, then, was a piece of original old-growth forest (later located a half-hour drive away in Pedevalasa) and I would have the makings of a comparative ecological field study on habitat use by birds.

I deployed thirteen mist-nets for four days in each habitat, and in each I also carried out eight half-hour morning censuses of the birdlife. Finally, I conducted a qualitative habitat-vegetation analysis of each plot where I worked. Vasantha Kumar, the assistant estate manager in nearby Lothugedda, provided the botanical expertise to identify the native trees. I was soon to be joined by Shahid Ali, who was a bird student and could help with the work.

The project instantly gave shape to my planned stay here in the hills. We could do our general bird surveys and some ecological research as well. I carried out additional studies of foraging by woodpeckers and mixed-species bird flocking. I was busy all the daylight hours with mist-nets, censuses, and general observations. I had found a gold mine for birds and bird observations! No word came from Ripley, and I learned that Ali would not arrive for another two weeks.

Each day had a routine. Up early. Observing and censusing. A hot breakfast with tea, plain omelette, and puri, the tasty hand-made fried flatbread that was synonymous with breakfast in southern India. Out to check the nets, continue observations, and range about in the various habitats, following my nose. Then lunch, a brief nap if the rains came on, afternoon observations, and closing of the nets. It was a marvelous way to spend a day—busy, but productive. Life was good at bucolic Wangasara.

On September 29, Shahid Ali arrived to join the field team (notably without the Ripleys or Sálim Ali). Grand-nephew Shahid had been studying the Indian Partridge at Point Calimere, in southernmost India. He was diminutive, erudite, and amusing, one of a new generation of urbane and educated Indian intellectuals. He was a fine companion as the days played out in the Eastern Ghats. We shared duties and enjoyed being out in the field in this spot that was new to both of us. Our host, Krishna Raju, was nowhere to be seen. Apparently he was tied down by his office work, and perhaps not a little disappointed that his much-anticipated trip for Ripley and Ali was not proceeding as expected (with two second-stringers instead).

South India was nothing if not the archetypal demonstration of how humankind and nature interact. Humans dominate the landscape, even in the rural areas. Here agriculture was king—plantation agriculture mixed with village cropping of legumes, grains, and vegetables. Expanses of production forest covered the hillsides, including large stands of bamboo, also an important production crop. Unlike New Guinea, little could be considered virgin land—humankind had grasped economic opportunity in all its forms. No surprise, given the pressing need and the population pressures. This, in all likelihood, was what most of New Guinea will come to look like in 2050 or perhaps 2100.

During the first six days of October we were beset by a large cyclonic storm—a typhoon—that came up the Bay of Bengal and dropped several feet of rain on southeastern India. It rained day and night all six days. We lost electricity in our little cabin. We lost access to the outside world when the plains below us were inundated and the roads were rendered impassable. Before the rains began, Appa Rao had gone off to Narsipatnam to buy groceries. He did not return for a week, because that town was under some six feet of water for several days. Our suffering was minimal, compared to the chaos that struck the plains and coast. But we did suffer. Our cook was gone. Our food supplies dwindled, as did our hope. We felt cut off from the world. No cars or buses passed by, and rain made the work difficult, at times impossible. It was a phenomenon I had never experienced in New Guinea, where cyclones are rare indeed.

Down on the plains, deforestation and habitat alteration had caused a catastrophe. The reduced forest on the ghat face was unable to hold the water that fell as rain, and it rushed down at high speed onto the plain. On the deforested plain, the land's ability to slow the movement of rainwater was minimal. The combination was deadly: rushing water, overflowing rivers heavily laden with silt and debris traveling at high speeds, improperly managed watercourses, engineering snafus, and the like. Dozens were killed, and thousands were rendered homeless by the deep water. India, the land of too many poor people and too little natural habitat, is a place where natural disasters are commonplace. The combination of abundant humankind and overtaxed natural systems creates the problem.

With no telephone, no radio, no newspaper, and no visitors, we were completely out of touch with the rest of the world for a week. We really did not even know what was happening down the hill. Food was short, but we continued our work during periods when the rain subsided.

So, where were Dillon and Mary Ripley? This was their expedition, after all. I had been sent over simply to get things set up. They were delayed because of the press of Smithsonian business. It was not clear when they would arrive. Shahid and I thought they might arrive any day with Krishna Raju, but that did not happen. It turned out that the Ripleys had to cancel and could not let us know until relatively late in our trip. It was a pity, for the two of us were eager to spend time in the field with Dr. and Mrs. Ripley, but that experience would have to wait until 1985. In any event, the change of plan allowed Shahid and me to get our footing in a new environment while learning the ropes in the backcountry of south India.

On October 13, after we had recovered from the cyclone's effects, we shifted to a remnant forest site named Pedevalasa, about a half-hour drive from Wangasara. We set up a tent camp and worked the remnant old-growth forest as well as the heavily logged production forest that surrounded it. It was quite a change from Wangasara—no nearby village, no rest house, no electricity, no plantations of teak, coffee, or pepper. Not even any gardens. It was more like New Guinea in many ways. Here wildlife was certainly more abundant. Macaques and Common Langurs were all around, as were Indian Giant Squirrels.

On my first walk around the area, I even had an exciting herpetological encounter. I was in an overgrown field, birding, in the late afternoon. As I watched a Rufous Woodpecker in the nearby forest edge, I heard a lisping-hissing that I initially took to be some bees visiting flowers. But no, it was a snake coiled atop a large shock of wind-thrown grass, its upper body rearing up with its neck broadened and arched in the classic spoon-shaped threat pose of an Indian Cobra. Because of its raised perch, its head was about at the level of my head. I watched it at leisure for a few minutes, and it never retreated or lowered its hood.

The Indian Giant Squirrel, a creature perhaps twice the size of the Eastern Gray Squirrel I know from the United States, was a rich reddish brown above and golden ocher below, its tail a huge bushy maroon appendage that flicked nervously. The first one I encountered was high in a tree feeding on the canopy twigs. Hanging by its hind feet, it would reach down and snap off a 4-inch section of twig, hold it in its paws like a corncob, gnaw off the bark, peel off the outer wood, then consume the inner wood of the twig. When it noticed me watching, it gave a colossal alarm call, a squealing, explosive *squoit squoit squoit*—not your average squirrel.

The days passed quickly at the Pedevalasa camp, as we censused and netted in the old-growth forest and did general observation throughout the area. The Common Langurs, with silvery white fur and black faces, were like grizzled, wiry old men. They moved in troops along the forest edge. Far more attractive than the rather raffish, scruffy macaques that haunted the edges of cities and towns, the langurs were a special treat, always around, but invariably wary and retiring. The primates in India were much more visible than those in Panama.

Once, a Brown Wood Owl appeared on an open perch at dusk in the clearing, transfixing me with its hooting. More remarkable was the discovery of a Ruby-cheeked Sunbird in the canopy trees near camp on two successive days. A widespread species in Southeast Asia, in India this bird had heretofore only been recorded in the far northeast; this was a first for peninsular India. We were now beginning to make original contributions to Indian ornithology. Krishna Raju had only the year before had a similar achievement by discovering Abbott's Babbler in this very place. Ripley and I subsequently named this new peninsular subspecies in Krishna Raju's honor.

We had a sleepless night at Pedevalasa because of a "tiger scare"—the men were certain a tiger was prowling the verges of our encampment. I had

Sálim Ali, the dean of Indian ornithologists,
in Andhra Pradesh in 1983

heard a lot about tigers, but there seemed to be more fiction than fact attached to the stories. Without question, our local helpers were terrified of
this powerful beast, present or not.

We shifted our camp to Lankapakalu, another highland area heavily planted
in shade coffee. There we were to meet up with Shahid's great-uncle Sálim,
the grand old man of Indian ornithology. I was filled with anticipation. In the
tiny hamlet of Lankapakalu we were housed in an unoccupied mud house
that was, in fact, quite comfortable—far more luxurious than our tent camp in
Pedevalasa. Lankapakalu, too, was more developed than Pedevalasa, because
of the hamlet and the active coffee plantings. The birdlife was satisfactory,
though there was no sense of wilderness here.

I was now getting the hang of being in India and was adjusting to all the novel happenings: the endless tea breaks, the luxurious breakfasts, the spicy dinners, the continual attention of assistants and attendants. It certainly did make it easier to focus my energy on bird research!

Sálim Ali arrived, causing as much local excitement as a movie star. This somewhat ancient but spry and tiny man was a marvel. Although eighty-seven years old, he was still active and restless, but also unassuming and polite. Shahid was clearly in awe of his uncle, as was I. Our common currency of discussion, of course, was birds, and there we could communicate as fellow students of nature. Although hard of hearing, Dr. Ali asked us lots of questions and insisted on going out birding whenever possible. He was an excellent field observer, and we all felt we were part of a rare experience being in the field with him, in spite of our disappointment about the Ripleys.

The first night Dr. Ali was with us, we had another bit of excitement. I was out back brushing my teeth in the darkness when a spine-chilling sound rent the night. It was my first *chitra-pooli* (Leopard). The beast was not far from the edge of the hamlet, giving a loud rasping moan, in and out, thirty or forty times in succession. Of course, at that moment I could only guess what it was. Dr. Ali came out and identified it for us. Hearing it was almost as good as seeing it. In fact, hearing it call from the verge of tiny Lankapakalu gave me goosebumps. I moved my little *charpoi* out to the walled backyard so I could sleep outside and listen for its horrific moans throughout the night. The Leopard did call out several more times. In response, villagers went into their gardens, banging pans loudly and shouting in order to drive the beast away. Their primary concern was their domestic animals, which were taken in numbers every year by these stealthy and resourceful predators.

Another treat was to follow: an early-morning mixed-bird flock, with Sálim Ali trying to enumerate all the species moving through the vegetation. We were particularly taken by the foraging of the diminutive Indian Piculet, a wren-sized woodpecker that on that day spent most of its time tapping lightly on tiny hollow stems in order to drive guarding ants out of their stem-hole nests. Upon emerging, the ants were sucked up by the piculet's sticky tongue.

Watching the piculet with the Great Man was a memorable treat—both were diminutive, interesting to behold, and rather rare.

Dr. Ali did not have much time to spend in the Eastern Ghats. Shahid and I showed him Wangasara and Lammasinghi, then saw him off at the Visakhapatnam Airport, bound for Bombay. Not long afterward, we wrapped up our fieldwork, thanked Krishna Raju, and headed home, satisfied that our first sojourn in southern India had provided lots of memories and the data for a scientific paper on birds in human-managed habitats.

In 1985 I returned to Andhra Pradesh to continue our collaborative survey of the birdlife of the Eastern Ghats. As before, I arrived in advance of the team, but on this occasion I was no longer a novice and knew better how to move ahead. I pressed Raju and the Forestry officials on where the richest and most interesting forest patch was and was told we should try Jyothimamidi. But we would need to steer clear of the Marxist rebels who ranged through those forests. For this expedition, I was joined by P. B. Shekar from Bombay, and by S. S. Saha and C. K. Mishra from the Zoological Survey of India in Calcutta. We would establish a field camp, and then the Ripleys would come in and join us. (Ali was not available for this expedition.)

Following the same slow and methodical steps as on the first field trip, our team made its way to the fabled Jyothimamidi (which in Telugu means "twin mango trees"). Nestled in a valley on the southwestern verge of the Visakhapatnam ghats, and at 1,500-feet elevation, Jyothimamidi was lower than Wangasara. In terms of wilderness in southeastern India, it was in essence at the end of the road—as far as one could drive by jeep into the wilds of the ghats. The road had been installed to get timber out of this rich valley, so we were working an area that had been selectively logged a decade before. It was still forest, but with clearings and some heavily thinned tracts.

That said, it was the richest forest I ever worked in peninsular India. We arrived in the late morning and immediately began to search for a campsite. Before long it became evident that one site was far better than any other; it lay in a cool glade in a small stream-bend with ready access to the road.

It had considerable flat ground for the tents, water for bathing and washing clothes and dishes, and a thick forest canopy that would protect our camp and the tents from the intense tropical sun.

There was just one problem. The local forester and his tribal informant told us that this particular spot was occupied by a Tigress and her two cubs. We were assured that we could not camp here. If we did, our team would come to grief.

I did not see any alternative, so I stood my ground and requested that the camp be placed right by the stream. Our crew of about fifteen (there is always an abundance of helpers in India) grumbled loudly, but they unloaded the jeeps and trailers and began setting up camp. No more than fifteen minutes elapsed before we heard the Tigress give several loud coughing barks a couple of hundred feet off in the thick brush. The helpers were much put out, but a small group of naturalists (including me) quickly grabbed binoculars and cameras and set off in the direction of the Tiger. The only track led through tall grass, and we quickly realized that we might be putting ourselves at risk. We slowed our pace. The Tigress barked again down the track, and immediately afterward we heard two local men down the track shout loudly in response. We all jumped, and decided to retreat back to the encampment. Barking Deer and squirrels sent up alarm calls in response.

We had displaced a Tigress and cubs for our camp. The big cat protested several more times that afternoon, and we were a bit nervous as dusk settled on the verdant site. Our Indian partners suggested that we burn our kerosene pressure lanterns through the night to keep the cat at bay (supposedly big cats do not like to approach well-lit camps). This we did. Our assistants refused to sleep in their appointed tents, but instead piled into the two jeeps to sleep in relative safety. I slept well until the lanterns began to run low of fuel and started to make a popping sound every minute or so. I had to drag myself from the comfort of my tent to manually switch them all off. The Tigress was nowhere to be seen.

We had five days of exploration and survey before the Ripleys were to arrive. Tracks led through the forest in various directions. Streams wound to and fro. Birds and mammals were everywhere. A superb adult male Peacock woke us the first morning, calling from a canopy tree above our tent site. The morning sun made the bird's iridescent feathers glitter green and blue. This was a natural paradise, enriched by the knowledge that we were treading pathways used by Tigers each night. We saw the places where the big cats scent-marked and scratched. We could feel them looking at us from their daytime hiding places. It was remarkable—a combination of elation and fear that heightened the senses.

We spotlighted for mammals and night-birds each evening. The forest supported Sambar and Barking Deer, Mouse-Deer, and Indian Porcupine, all of which showed up in our spotlights with regularity. The area was rich with game. One night Saha caught some eyeshine low to the ground. He whispered "Fishing Owl" and we all put our lights on the site; two of us put our binoculars onto the creature, which on better examination was an adult Leopard sprawled on a large rock. The Leopard did not seem to mind the fuss and remained on its perch for several minutes while we took turns marveling at it. Then it was gone in the night.

Being a woodpecker lover, I was in heaven at Jyothimamidi. It was the richest site in all of India for woodpeckers, with eleven species: Speckled Piculet, Rufous Woodpecker, Large Yellownaped Woodpecker, Small Yellownaped Woodpecker, Himalayan Goldenbacked Three-toed Woodpecker, Great Black Woodpecker, Fulvousbreasted Woodpecker, Yellowfronted Pied Woodpecker, Browncrested Pygmy Woodpecker, Heartspotted Woodpecker, and Larger Goldenbacked Woodpecker. Woodpeckers were everywhere! They joined mixed-bird flocks. They foraged for nectar at flowering trees. They drummed incessantly. We collected a specimen of the Himalayan Goldenbacked Three-toed Woodpecker—the first south of the Himalayas. We also made the first sight record of the Great Black Woodpecker for Andhra Pradesh state.

Our favorite was the endearing Heartspotted Woodpecker, the most adorable little woodpecker on earth. Harlequin patterned in black and creamy white, it had a wonderful recurved crest that gave it a particularly saucy look. And its back had a series of little heart-shaped spots, for an elegant look that could have been designed by Martha Stewart. We would encounter pairs of this perky woodpecker in the early morning, perched atop a spire of dead tree in the forest. Pairs of this uncommonly encountered species would move nervously about on the spire, emitting a high sweet series of piping notes for minutes at a time. Their performance was a highlight of our expedition.

The only nuisance at Jyothimamidi were the seed ticks, tiny chigger-like bloodsuckers that populated this game-rich forest in abundance. We each got hundreds of seed-tick bites, and the itching they produced was nasty!

❧❧

At last the Ripleys arrived, trailed by a retinue of Indian bureaucrats from state and national agencies. We got them set up in their large tent and made our best effort to disperse the bureaucrats. As they departed, they all warned us about the Naxalite (Marxist) rebels who infested this backwater region of Andhra Pradesh. They were concerned that our party would be kidnapped for ransom. The Ripleys found this amusing, but were more concerned about the threat of the Tiger.

All attention was now on the Ripleys, making certain they were comfortable and properly attended. Our field survey effort dropped considerably. The Ripleys were in their seventies at this point, an age when few would consider camping in a South Indian forest infested by Naxalites, Tigers, and seed ticks.

The Ripleys were entertained by being here at Jyothimamidi, but they were no longer really "field ready." It took me a while to realize that. Each morning I rose well before dawn and went out to walk the forest trails in search of birds. I would return by seven-thirty for a quick omelette breakfast, make notes in my diary, then go out for a second survey. Upon returning at about ten, I would find the Ripleys conducting their morning ablutions, preparing for breakfast. As the heat of the day built, the Ripleys would be preparing for their daily excursion in search of birdlife. They would be gone from camp from eleven until about twelve-thirty. Upon returning, Dr. Ripley would complain that they saw virtually nothing on their jaunt. This happened repeatedly.

Before long, Dr. Ripley announced that it was time to move to a better site. He was sure this site was very poor. Our team tried to make clear that it

S. Dillon Ripley (*right*) in conversation with the author in Tamil Nadu, 1985,
with Mary Ripley and C. K. Mishra in the background (photo by S. S. Saha)

was the very best Andhra Pradesh had to offer. Our perceptions differed because we were out early, ranging far and wide, and the Ripleys, given their age
and condition, went out only briefly in the heat of the midday, when there was
no wildlife to see. I came to understand that field natural history is a young
person's pursuit. The Ripleys were simply beyond that stage in their lives.
They wanted to do the job, but their complicated lifestyle made it difficult.
Sálim Ali was the exception that proved the rule. His spartan bachelor lifestyle, ruled by single-minded ornithological focus and asceticism, endowed
him with a young man's habits. The Ripleys, having for decades operated
as intellectual leaders of the Smithsonian Institution and all it entailed, had
naturally drifted from that spartan regimen.

One morning I returned to Tigress Camp to find the Ripleys in discussion with a new delegation. Not knowing who they were, I asked one of our
team and was quietly informed that the local head of the Naxalite rebels was
here with some of his entourage. The discussions seemed to go well. The
Ripleys were the perfect hosts, and the Naxalite party departed all smiles and
waves! We were not kidnapped, but the Ripleys decided it was time to move
on. So our team, saddened by the knowledge that it would be downhill from
here, packed up and moved to the first of several lesser sites.

The three subsequent campsites we occupied could not hold a candle to the Jyothimamidi forest. In addition, the seasons were turning and the heat of Indian spring was upon us. The sun roasted our camps. Bird surveys in India are best conducted in the cool of winter, when the resident birds are joined by the Himalayan migrants. At our last encampment at Anantagiri, the only forest patches we could find were in little ravines beside the ever-present coffee plantations (mainly sun coffee). At that point, we thought back wistfully to the cool waters of our Tigress Camp at Jyothimamidi.

In 1986 I was back in southern India. With the Ripleys I visited the High Wavy Mountains in Tamil Nadu (very close to Periyar Lake in Kerala). There we stayed in a tea estate in the high wet uplands, where we had rain and forest leeches in abundance. From the edge of the high plateau I could gaze from the step scarp face into the Kambam Valley to a lowland forest tract that seemed remarkably undisturbed. Looking down more than a thousand feet, from time to time I could glimpse the black-and-white forms of Great Indian Hornbills winging over the canopy. I decided that this was the place to set up a survey camp after the High Wavies. It would be a bracing change from our rather posh setup in the tea estate's rest house, with an ancient house servant who served us three sumptuous meals a day, perhaps reminiscent of the Raj era. Doing bird surveys from a well-provisioned estate rest house was fine, but it was not jungle living!

After the Ripleys departed for the United States, our field team spent several days searching by car for that mysterious lowland forest. Virtually every track we drove down was littered with the brown softball-sized dung left by Indian Elephants. Biswanath Roy, my colleague from the Zoological Society of India, was not at all pleased to be driving down tiny single-lane paths where around any bend we might come up against a "lone tusker" who could instantly overturn our car in a fit of pique. Elephants kill people by the score every year in India.

After interviewing some local forest officials, we were pointed in the direction of the Vannathiparai Reserve Forest. At the southern terminus of the spectacular Kambam Valley, set against the western wall of the valley, lay this forgotten forest. Most naturalists visiting India's far south are drawn to the world-famous Periyar Sanctuary. Nearby Vannathiparai was better, though finding it was not easy. We had to wander through low sandy scrub to find the entrance track. We parked in a parched sand barren, occupied by nothing but thornbushes. From the parking spot, no forest was visible. I was not hopeful as we moved down the sandy track, but we soon entered some woods, which quickly rose to be a majestic lowland jungle forest, watered by a lovely clear-water stream. Here at the streamside Roy and I agreed we would camp.

Our forest guard was concerned. He told us it would not be safe to camp here because of the elephants. We had a conversation similar to the one at Tigress Camp in Jyothimamidi. Roy and I prevailed, agreeing that we would hire night watchmen who would tend two bonfires on either side of the camp to keep the elephants at a distance.

Over the next day we transported our bulky camp gear to Vannathiparai Forest. We put up our big nylon tents, set up a screen work tent, and constructed a covered dining area in the open forest interior beside the stream. We were joined by several locals from the hamlet a mile downstream. These men were to tend the bonfires and serve as the night watchmen. We did not know it at the time, but we were camped in a tiny remnant of South India's natural history past—a small patch of old-growth forest that supported the original Indian megafauna that today is on its way to oblivion. Over the next three weeks, while we surveyed birds and mammals, we would be looking back at what the Indian subcontinent was like before the advent of eight hundred million people.

Our first night was instructive. It got dark around six-thirty. Roy was organizing our first dinner, and the watchmen were setting their bonfires (which I thought ludicrous). Suddenly, the evening's peace was broken by a horrific din a couple of hundred yards upstream: the discordant trumpeting of elephants that had been displaced by our camp. The forest guard was right. This was, in fact, the regular evening watering hole for a herd of Indian Elephants that inhabited this forest. They were emphatically announcing their annoyance with us, the interlopers. The men hastened to light their bonfires and shouted back at the elephants to keep them at bay. Roy assured us he would stay up all night with the twelve-gauge shotgun to protect us.

The next morning, after a night's sleep undisturbed by the trampling herd of elephants, I went out early in search of birds. Following a jungle path, I chased down a Greater Goldenbacked Woodpecker—like a garish version of our Pileated. Watching this red-crested woodpecker in a tree across a tiny forest clearing, my eye was drawn to a white object low in the clearing. It was the lower foreleg of the largest of the wild cattle, a bull Indian Gaur, foraging peacefully in the clearing and oblivious to my arrival. This massive beast was no more than 50 feet from me, browsing on the leaves of a sapling and virtually invisible in the dappled light of the forest. In size, it is second only to the Indian Elephant in Asia. The Indian Gaur stands about 6 feet at its massive shoulder. Its glossy coat is black, with offsetting white on the lower legs. A golden forehead is bracketed by an impressive pair of curving pale horns. This is the great forgotten beast of the Indian forest, living in the shadow of the Tiger and Indian Elephant. An old bull can weigh a ton. After a few minutes it realized it was being watched by a human and instantly vanished into the green undergrowth without snort or sound of any kind.

The forest was infested with troops of Long-tailed Macaques. We saw them many times a day, and the troops were quite tame. Individuals foraged on the ground and little ones gamboled in the low branches and vines. I spent a lot of time watching these smallish dun-colored primates. They shared the forest with the frosty gray Indian Langurs, which were much more wary. Indian Giant Squirrels were common in the treetops.

As several days passed, I learned the trail system better and better and was able to range farther from camp in search of wildlife. One track in particular led up a ravine onto the face of the High Wavy escarpment. Picking my way down the track in the late afternoon, I looked up to see a group of five rusty-colored dogs, loping through a rocky ravine bottom. I trained my binoculars on them and noted a bushy, dark-tipped tail and a general color pattern that closely resembled a Red Fox. This was a pack of Dholes, also known as Indian Wild Dogs. This predaceous Asian canine shares the Indian peninsular forests with the Wolf, Jackal, and two fox species. Dholes have a somewhat fearsome reputation, but on this day they seemed to me to be quite beautiful and nonthreatening.

Upon returning to camp, I told everyone of my Dhole encounter, which drew murmurs of jealousy. But then I was interrupted by the arrival of one of the watchmen, who shouted something in Malayalam and showed himself to be quite agitated. Roy translated that the man had gone out to collect firewood, but the main track had been blocked by two enormous King Cobras who were "wrestling" in the path. He had waited ten minutes, but they remained engaged so he had returned in fear. We rushed out with him to see the cobras making love, but by the time we got to the site the two snakes were nowhere to be seen. The watchmen said the snakes were more than 12 feet long and stretched across the entire path.

The elephants complained for two consecutive nights, then apparently moved to another watering hole in the forest. We focused on birds, and we were not disappointed. There were hornbills, leafbirds, frogmouths, parakeets, sunbirds, coucals, woodshrikes, trogons, junglefowl, paradise flycatchers, treepies, minivets and fairy bluebirds. In southern India the birds are not invisible as in New Guinea; they are out in the open. There is no place quite like South India for watching birds.

The most interesting story to be told of the Eastern and Western Ghats of southern India is the relictual distribution of the humid forest birds. Whereas the birds of the Indian peninsular plains are mainly of Indian origin, those of the Indian hills share their affinities with the eastern Himalayas and Southeast Asia, far to the northeast and isolated by stretches of lowland plains in Bihar and Bengal. Thus the Ruby-cheeked Sunbird that I discovered at Pedevalasa is absent from the rest of peninsular India, but is widespread in Assam and Burma and elsewhere in Southeast Asia. Such relict species inhabit only the forested hills and mountains in South India, but occur in lowlands and hills in Southeast Asia.

What earth processes caused this scattered and relictual Indian distribution? It seems that a past cycle of climate change most likely produced this pattern, repeated in scores of forest-dwelling species of birds and mammals. Apparently in the past there was a period of warm and wet climate that must have fostered the development of humid forest in a continuous band from southern India northeastward to the eastern Himalayas, and thence eastward and southeastward into Burma and Thailand. More recently, a South Asian dry cycle has led to the retreat of humid forest from the Indian plains, with the remnants holding on in the Eastern and Western Ghats. These areas of relief produce local concentrations of annual precipitation that can support humid forest. The birds and mammals dependent on humid forests retreated with the forests into the nooks and crannies of the ghats, where they hang on today. Thus the ghats forests are relictual, with remnant biotas of a former richer and wetter India.

Similarly, the Indian megafauna (the Indian Rhinoceros, Indian Elephant, Tiger, Gaur, Indian Lion, Leopard, Nilgai, and Blackbuck) are on the brink. A once-rich and oversized mammalian fauna is presumably awaiting the next extinction pulse to take it to the land of never-more. India's Cheetah population disappeared in 1948. India's Lion population is down to about three hundred. Some eighteen hundred Tigers haunt India's remnant forests. Which species will be next to go?

One obvious reason India's megavertebrates are on the way out is the abundant instances of conflict between the wild mammals and the burgeoning human population. Before the advent of humans, Indian Elephants carried out regular annual migrations in search of suitable forage. The elephants still attempt to carry out these annual movements, but it becomes more and more difficult because of the barriers created by human development. In the High Wavy Mountains, the tea estate manager told us about a herd of elephants that migrated through the estate every year. It followed the exact same path each year. When a wall was built in a garden that transected the traditional route, the elephants walked their traditional path up to the wall, their memories guiding them, wall or no. In some instance, elephants push down such a wall, to avoid diverging from their traditional route. Naturally, elephants knocking down walls present a problem to landowners and estate managers. Elephants trample crops and attack people and even stomp down village huts in fits of anger and frustration. Tigers take down woodcutters and forest workers. Leopards collect individuals sleeping on charpois on their front porches. Cobras strike and kill unsuspecting people who mistakenly tread on them.

Unless these points of conflict are actively resolved, Indian societies will not make way for the animal life sharing the environment. It is the task of the environmentalist to establish rules and procedures to reduce conflict, and to allow human and animal societies to coexist in what remains of India's natural environments. The challenge becomes greater every year, as the human population grows and as the natural habitat becomes ever more fragmented.

On these three journeys to India, I saw a civilization of great antiquity and a land where abundant people share space with a remarkable megafauna. A far cry from New Guinea! In India I started to lose my New Guinea chauvinism and, I hope, became a little less sheltered and naive about the world. I learned that I could operate in a South Asian society and appreciate its particular pleasures and adjust to its nuisances. I also began to appreciate that each corner of the world was home to magnificent faunas as well as fascinating humans. The "tribals" who peopled the Eastern Ghats were easily as interesting as the clans I knew from Papua New Guinea, though less masters of their own world. These were my first visits to one of the earth's impressive continental faunas, and though I could see the dark threats pressing at the door of India's natural world, I also could see that even when threatened and not a little degraded, the wilds of India had much to teach a naturalist.

What does New Guinea have to learn from India? A lot, I believe. For in India we can see the by-products of the long presence of complex and ancient human societies and the impacts of rapid population growth. Looking at India today is probably equivalent to looking at New Guinea in the year 2050 or 2100. New Guinea has much to lose by following India's developmental pathway of high population and intensive agricultural development. In India, the loss of forest has brought about human misery of various forms: shortages of clean drinking water, catastrophic losses of topsoil, deadly flooding, persistent local famine. These are rare today in New Guinea, but with forest clearance all could become commonplace. The poverty so widespread in India is still relatively little known in New Guinea, mainly because of traditional land tenure and abundant land and agricultural potential. We can see how precious are the resources New Guinea rural communities possess in abundance: arable land equitably distributed, abundant and clean water, forest and forest products, game and fish for protein. The people of New Guinea need to study the developmental histories of neighboring nations in order to determine which practices will best ensure their well-being for the long term. It is the duty of its leadership to make choices based on this sort of wise research.

What one can see happening in India, and what one is beginning to see in Papua New Guinea, is the westernization and urbanization of the societies. People look toward the cities rather than the forests for economic answers. The cities offer cash employment and a predictable lifestyle and health care. At issue, however, is the loss of human appreciation for the critical services provided to society by natural forested ecosystems. Ironically, the only truly self-sustaining lifestyle is that of forest-dependent swidden (slash-and-burn) agriculturalists living in small family or hamlet groupings. In this swidden lifestyle, the forest provides, year after year, fish, game, and housing materials. The forest provides a reliable supply of clear drinking water. And the forest ensures the abundance of rich agricultural soils through the swidden cycle, in which garden plots are left fallow after several years of crops, to be recleared in future decades once the soil is replenished by the regrowing forest.

I am not advocating that we consider reverting to a forest-dwelling swidden lifestyle. Rather, I believe we must examine how the natural world produces so many of society's goods and services. We need to take steps to allow the Earth to produce these goods and services (which it does for free) in as many places as possible, and in ways such that urban and suburban areas can obtain maximum economic and lifestyle benefit from them. Natural resource economists can calculate the full cost of the various developmental pathways. Governments need to take account of the full costs of urbanization and make choices to reduce the impact on our local and regional ecosystems, if we wish to support a successful lifestyle for the twenty-first century and beyond.

CHAPTER FOUR

Wallace's Promised Land

I looked with intense interest on the rugged

mountains, retreating ridge behind ridge into the

interior, where the foot of civilized man had never

trod. There was the country of the cassowary

and the tree-kangaroo, and those dark forests

produced the most extraordinary and the most

beautiful of the feathered inhabitants of the earth—

the varied species of Birds of Paradise.

Alfred Russel Wallace,

The Malay Archipelago (1869)

ALFRED RUSSEL WALLACE, who with Charles Darwin first discerned that life on earth is the product of natural selection and the evolutionary process, made these remarkable deductions during his long sojourn in the Malay Archipelago in the 1850s. A high point of this world-changing natural-history trek was Wallace's visit to westernmost New Guinea—for him the mystical land of the bird of paradise. Even today, Wallace's spare and clear-sighted field accounts are good reading. He was undoubtedly one of the world's great field naturalists. Every naturalist who ventures to the western side of New Guinea follows in Wallace's footsteps.

The western half of the large island of New Guinea is easternmost Indonesia. Today we call it Papua or West Papua, but over the years it has had many different names, reflecting changing colonial and postcolonial conditions. For a century it was part of the Dutch East Indies. In the 1950s it was Dutch New Guinea. After the Indonesians took it by force from the Dutch in 1963, it was named Irian Barat (West Irian). By 1981 it was Irian Jaya (victorious Irian)—a world virtually lost to tropical biologists since the Indonesians had taken control nearly two decades earlier.

Irian Jaya was, of course, a lot like Papua New Guinea, but wilder, less populous, and with higher mountains and many more barriers to exploration by outsiders. The Suharto regime held a tight rein over Irian Jaya and did not like foreigners nosing around. The military had taken a number of "police actions" against the small band of Papuan freedom fighters (the OPM). But because of its isolation and little-studied biodiversity, it was a highly desirable destination for a footloose field naturalist.

In September 1981, I met up with Brian Finch in Papua New Guinea, where he was based in Port Moresby. Finch was a passionate birder and naturalist. He and I jetted down to Cairns, in northern Queensland, to pick up a charter aircraft to Timika, the town that served as gateway to the great Freeport gold mine in Irian Jaya. We had solicited an invitation from Freeport through the connections of my new boss at the Smithsonian, Dillon Ripley.

Opposite: Clouds and mountains, Irian Jaya

When we inquired in the Freeport office at Cairns about the charter, we were told that because of overbooking the two of us could not fly until the following week. The delay was a disappointment, but it did give us an opportunity to look around the back country of northern Queensland.

First Brian called his friend Ian Mason, a biologist with the Commonwealth Scientific and Industrial Research Organization (known simply as CSIRO). Mason made the hour-long drive from Atherton to fetch us and generously put us up in his home. Mason then dropped everything and took us birding for a day and a half.

Although the rainforest plants and animals of northern Queensland are similar to those of New Guinea, I was now to see the quite noticeable differences of Australia's other dominant habitats. Australia's abundant open woodlands and savannas are dominated by eucalypts and melaleucas and are not nearly as rich in species. Ground-dwelling wallabies and kangaroos, of course, are much more commonplace. Open-country parrots (galahs, rosellas) are prototypically Australian—absent from New Guinea. In general, Australia is dry and temperate, with patches of wet rainforest, whereas New Guinea is wet and tropical, with patches of dry-zone woodlands and savanna.

Ian showed us around the Atherton Tableland. We visited Lake Barrine National Park, where we birded in the rainforest and had tea and scones at lakeside. The highlights of our visit were the giant Kauri (*Agathis*) conifers with trunks 10 feet in diameter, and a displaying Victoria's Riflebird, one of Australia's endemic birds of paradise. The giant trees were among the very few that had survived the logging of the early twentieth century. The riflebird, a velvety-black apparition, performed a wing-waving dance atop a tree stub 20 feet above the ground, right beside the trail. In New Guinea, the Magnificent Riflebird gives a similar dance, but finding a site for it is extremely difficult because the bird is incredibly shy and its dance ground is hidden deep in the forest.

We also visited Curtain Fig, a giant strangler fig that through some weird quirk had created a wall of roots like a massive curtain in the rainforest. The Crater, another birding stop, was an ancient, water-filled volcanic caldera in the rainforest. Here we encountered the Tooth-billed Catbird, which lays leaves on the ground as part of its display, and the Australian Brush-Turkey, a black turkey-like mound-builder splashed with yellow and red on the bare skin of its face and wattled neck. In Australia, this brush-turkey was a tame denizen of the Crater's parking lot, hunting food in the scattered leaves covering the gravel. In New Guinea, such a succulent piece of game would be hunted to rarity. Culturally, Australia today is a world apart from its northern neighbor.

Australia's Atherton Tableland is, without doubt, one of the most pleasant corners of a continent that is bursting with beautiful natural destinations. It is a fertile and verdant rolling upland plateau ringed in rainforest, but with attractive agricultural communities in its center. I have since visited several times, and every time I come away thinking "This is where I would like to live."

What about the Tableland is so special? Certainly, to a New Guinea naturalist there is the lure of rainforest without the fuss and muss. On the Atherton Tableland, I could live in a well-provisioned, peaceful, modern small town and work in rainforest only a short drive down a paved road. In New Guinea, access to the forest in most cases involves a considerable hike or a flight in a bush plane, because near the towns the native vegetation has been removed for plantation crops.

While major development has taken place on the Tableland, the government has set aside beautiful and accessible reserves with well-maintained trails, explanatory signs, and general Australian good cheer. Australians are, without doubt, the most friendly and good-natured of peoples. Witness the response of Ian Mason, a busy professional biologist, when Brian Finch called him out of the blue. It was nothing for Mason to host us for a couple of days, birding and traipsing about, sharing his favorite natural places. Ian put his home, his refrigerator, his car, and his expertise at our disposal—all for camaraderie and friendship. This is the Australian magic, where everyone calls you by your first name ("G'day, Bruce!").

❧❧

The second time around, we managed to get seats aboard a decrepit Fokker F-27, filled with Freeport staff, on its three-hour flight to Timika, the lowland base of operations for the Freeport mine. The flight was fascinating, up the back side of Cape York Peninsula to Weipa, an aboriginal community. We touched down there for twenty minutes. Weipa is perhaps most famous for its bauxite, which is mined by scraping the top layer of bauxite-rich soils from the land's surface. For me, the highlight of that brief visit to this rather desolate spot was the sighting of an Australian Bustard at the end of the airstrip. It flew up from the verge of the blacktop with its broad wings vividly patterned in black and white. One of the most remarkable families of open-country birds because of their behavior, plumage, and stately demeanor, bustards inhabit Australia, Asia, Africa, and Europe. Today most are rare or declining because of the destruction and conversion of natural grasslands around the world.

The second leg of our charter flight took us across the Arafura Sea, making landfall at Dolok (Frederick Hendrik) Island on the belly of south-central New Guinea, just west of the border with Papua New Guinea. We crossed over a network of rivers and mangrove swamps, into a vast tropical wilderness sparsely inhabited by the Asmat. The swamp people are famous for their wood carvings that render the human form in an otherworldly and haunting fashion; they are considered among the finest carvings of Pacific traditional art. Next, we overflew a broad expanse of lowland alluvial forest, broken only by roiling rivers that came out of the cloud-shrouded Snow Mountains. Soon we could see an enclave carved from the vast green jungle, the centerpiece being a big concrete-paved airstrip. Here we put down, trundled out into the humid air of Timika, and waited to pass through the immigration barrier in a small wooden terminal open to the tropical air.

Although Timika currently is the "city" from which the mine is managed, in 1981 we did not even pause there. We crowded in a Chevy Suburban SUV and headed up to Tembagapura (Copper Town), the mine's nerve center. Here the mine staff lived and ventured out each day to operate the huge open pit mine atop the Ertsberg. This ore body sat atop a lesser summit a few miles from the highest peak in the tropical Pacific, where glaciers shone white and seasonal snows dusted the roof of New Guinea.

Timika is in the flat lowlands, Tembagapura 7,000 feet above. To

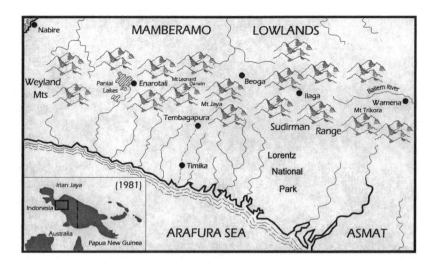

get there, the company has engineered a road up one of the steepest, most unstable, and rainiest mountain scarps on earth. The southern slope of the central range, New Guinea's highest and most rugged, is something to behold—basically a sheer wall that has been heavily dissected by rushing, boulder-laden torrents that dig deep and precipitous gorges in the uncon- solidated sediment. More than 250 inches of rain fall here annually, relent- lessly pounding these mountains and creating erosional havoc. Imagine the initial thoughts of the engineer assigned to build the auto road to the mine site! Year-round torrential rains, earthquakes and landslides, jungle-dwelling Amungme and Kamoro people who were essentially unknown. Deep ravines rumbling with angry water and tumbling rocks. Cloud cover that seemed never to permit the sun's drying rays to penetrate.

Instead of switchbacking like most European and American mountain roads, this gravel- and limestone-surfaced road took the traditional New Guinean route—up the ridge crest, grade-be-damned. After racing out of the alluvial zone, the road broke upward with a frantic fury; in the drizzle it almost seemed we were in a small plane lost in cloud. We passed rapidly through vegetation zones, then found we were overlooking deep torrent- cut crevices in the mountain wall, in one place plunging through a tunnel

where the ridge was too steep. After seventy minutes of fast-paced naviga-
tion we came to what looked rather like a prison camp set in a forbidding
mountain jungle (think *Bridge on the River Kwai,* but at high altitude). This
was Tembagapura. The high fence and guarded gate was necessary not to
keep prisoners in, but rather to keep the curious and incessantly scavenging
Amungme people out.

The Tembagapura camp was home to a rotating staff of engineers,
heavy-equipment operators, geologists, and the like. The men worked seven
days a week for thirty consecutive days, then took two weeks off in Australia.
Freeport had fashioned a town perched in a rugged little hanging valley
among the razor-edged ridges reaching up to the glaciers of Puncak Jaya
(Mount Carstensz), some 16,000 feet above sea level. This rain-soaked subtropical town was less than 5 miles, as the crow flies, from the remnant glaciers that cap these mighty mountains. Tembagapura is the twentieth-century
version of a gold-rush town in an inhospitable climate—a modern-day Yukon.
You can ship in the amenities (at great cost) but you can't really manufac-
ture civilization. The rawness, the mud and rock and rain, and the concentrated hormones of all those young mining men, mix to produce something
that is not quite a town, and indeed has more the look and feel of a prison
camp.

That said, Brian and I were glad to finally be here. Yes, it was ugly (how
could it not be?), but it was in the bosom of western New Guinea's highest
mountains and it had all the amenities that we needed—a road that traversed
from sea level to 11,000 feet, a twenty-four-hour cafeteria, our own dorm
room in the F Barracks with daily laundry service, even videos at night. We
settled in and met with our assigned host, the director of public relations. He
was an Australian missionary who had signed on with Freeport to manage
everything related to the public in this unlikely city in the clouds.

Our first afternoon, we decided to get out onto the road to have a
look at the birdlife and the forest. No natural vegetation was left in the camp
itself, so we simply walked out the main gate (open during the day) past the
ramshackle military post across the road, and up the road in search of birds.
In a few tantalizing minutes, before the heavy rain drove us indoors, we

saw a Black-mantled Goshawk soaring overhead, some Western Mountain White-eyes in a small *Pipturus* tree, and what may have been a Lorentz's Whistler (a rarely seen Western species), in forest-edge scrub. Pounding rain and thick mist closed in on the town for the rest of the day, confining us to quarters.

The next morning we were up before dawn, breakfasting in the mess, then out the gate again and up the road into the unknown. We knew of no ornithologist who had surveyed the avifauna of this cross-section of New Guinea. Although we expected no new species, we believed the little-known southern scarp of the range at high elevations might have some surprises. By 6:10 A.M. we were in a small patch of roadside forest in a narrow side ravine. The dripping, stunted vegetation offered up our first Short-tailed Paradigalla, a velvety black bird of paradise, which was probing aggressively for insects into moss on a tree branch. A small flock of Mountain Firetails searched for seeds in the roadside grass. At 6:20 we came upon a real treat: an adult male Splendid Astrapia bird of paradise, perched in the high dead branches of a tall roadside tree, giving a monotonous series of croaking notes. The display of this bird was undescribed, and it looked as if we had a male on a display perch. I had hopes of making a concentrated effort to nail down the behavior of this magnificent creature while here in Tembagapura.

We followed the road higher and higher, birding the whole way: Little Eagle, Brown Sicklebill, Spotted Berrypecker, Loria's Bird of Paradise, Orange-billed Lorikeet, Mottled Whistler. It was a paradise for little-known mountain-dwelling forest birds. After a couple of productive hours, we were high above the town and could look down on our godforsaken encampment, hidden below a tangle of ridges that guarded the heights of New Guinea's greatest summits. Looking up toward the mine site and the peaks, we could see little more than gloomy mists and thick cloud.

After an early lunch in camp, we visited the military post to present our credentials, accompanied by the Freeport mine public affairs officer. We showed our passports and visas, and made a statement to the neatly dressed military men there in the small, unimposing office. Then the officer in charge talked briefly with our host, and we departed. After crossing the

road and returning to camp, the public affairs officer primly told us that it
had been settled that Brian and I were not to be permitted to carry out any
ornithological surveys. The military staff had determined that it was not safe
for us to wander the road. Instead, we were to remain within the confines
of the barbed-wire perimeter fence until we could be flown out on the next
charter to Australia.

We were dumbstruck. Six months' effort up in smoke with a single
seemingly harmless discussion in the local language! We believed it might
have been punishment for our morning birding exploits up the road prior to
our formal request for a *surat jalan* (the official travel document allowing us
to be in Tembagapura). Our Indonesian visitor's visa, obtained in Australia,
only got us into the country; it did not confer any real status, once we were in
the military-managed backcountry of Irian. The all-important surat jalan had
eluded us! When we pressed the public affairs officer on how we could rectify
the situation, he simply told us that everything would be all right as long as
we spent the remaining six days within the fence and kept to ourselves.

We returned to the mess and chatted up several of the Freeport people
we had met the preceding evening—a geologist, a teacher, and an engineer.
They were surprised and regretful about our situation. We asked them how
we could seek some relief. Time was passing and we were not out birding.

Over the next day, a scheme was hatched and refined. We would talk
to a member of the Indonesian national staff who worked in the main office
and ask him to approach the military, with an offer from us to contract out an
armed escort to protect us as we birded the road for five days. That sounded
reasonable enough. It was worth the money and it might do the trick. So we
visited our office contact, presented him with the concept. He listened to
our plan coolly and told us to check back in an hour.

Within twenty minutes, we were accosted by the public affairs officer,
furious, telling us in no uncertain terms that we had broken our trust with
him, and that he wanted nothing to do with us for the remainder of our stay.
He turned on his heel and marched away, leaving us to fend for ourselves.

We were, in effect, placed on house arrest in F Barracks until we could
be offloaded onto the plane headed back to Australia. Here we were, in one

of the least-known corners of the least-known province of Indonesia, with everything but permission to do the one thing we wanted to do—get into the jungle and watch birds. Certainly, this was the low point for us. The rain bucketed down, mist and fog shrouded the grim valley, and all we could do was eat and read and watch the evening movie—and fret.

For what seemed like an eternity we sat, and slept, and ate, and stood on the porch of F Barracks and looked for any birdlife that might pass through the vegetation outside the back fence. We prowled the fenceline hungry for new observations. Just our luck: most of the original vegetation had been cleared away to make the camp, and only young thickets of saplings stood in the place of old-growth forest. Here we saw Oriental Cuckoo (a migrant), Sacred Kingfisher (a migrant), Pacific Swallow (a migrant), Torrent Lark, and Orange-breasted Pygmy-Parrot. Five species in four days. Out on the road we would have seen fifty species a day. We never again heard or saw the Splendid Astrapia, or the Loria Bird of Paradise, or the Short-tailed Paradigalla.

Instead we listened to the mealtime tales of science teacher Charles Prewitt and geologist Brian Partika. They were able to paint glorious pictures of the wonders we were unable to see for ourselves: Sunsets up at Wanigon Camp, where the Macgregor's Birds of Paradise pass by in small parties every day. Scenic Amungme villages in the forest east and west of here. Stories of fossils and new mineral strikes and life in this fascinating end-of-the-world location. We looked through the fence to see small bands of Amungme. They were gleaners, shadowing the camp in search of treasures discarded by this profligate society that was wildly rich in material possessions beyond these peoples' dreams. We watched the military with their M-16 rifles casually brandished about, as they killed time outside, smoking their clove-scented cigarettes during lulls in the rain. And we slept and dreamed of greener jungles, where we could hike, unfettered, and view the birds we had dreamed of seeing.

The geologists we talked to were remarkable. They were, like us, immersed in the wonder of the natural history of western New Guinea. Their passion was mineral formation and geological clues to the presence of concentrations of gold, copper, silver, and the like. They loved this work in Irian. While we were there, they were in the process of discovering the Grasberg lode—Freeport Indonesia's second and much bigger strike, just a stone's throw from the initial Ertsberg deposit. They had every technical resource at their fingertips, and they were doing what they had been trained to do under amazingly difficult conditions. One can appreciate the ardor these experts must have felt when they knew they were on the trail of a billion-dollar lode of gold.

These fellows appreciated our interest in birds. They saw lots of wildlife while in their field camps, and they could understand our passion for the rare wildlife of these rugged alpine habitats. We were both playing out our hunting-and-gathering fantasies, the inheritance of our distant ancestors.

Finally, after days and nights that dragged on interminably, the morning came for our departure. We loaded into a Freeport SUV and bumped down the road to Timika and its big airstrip hacked from the jungle. At that point we were no longer interested in toeing the line with Freeport. We wanted to see what birds we could in our waning hours in Irian Jaya. The moment we checked in at the airport and got our boarding passes, we took off down the road to the nearest forest patch and went to work. We could hear Greater Birds of Paradise cawing loudly from the forest canopy. We milked that patch until we heard the Fokker banking in. We were the last to board the plane, and that final hour of frantic birding produced sightings of the bird of paradise and other western lowland birds that helped ease our frustration, if only a tiny bit.

In 1987 I returned to Irian Jaya. Dr. Ripley and I were preparing to mount an expedition to the fabled Foja Mountains there and I was in Indonesia doing advance work. Jared Diamond had visited the Fojas in 1979 and in 1981 and had grabbed the world's attention by announcing rediscovery of the Golden-fronted Bowerbird. Lost to science for eighty years, the bird had been originally described in 1895 from specimens collected from an unknown locality.

Before heading to Irian, I met with a number of counterparts in Jakarta —government officials, ornithologists, long-term colleagues of Dr. Ripley— to build a consensus for this expedition. I traveled to Jayapura to consult with officials of the provincial government, the university (our proposed in-country sponsor), and the nongovernmental sector. In addition, I met with a number of mission groups; our work would need a plane and a helicopter, and the missions were in the best position to provide these in localities relatively close to the Fojas.

My thinking at that point was that we could stage the expedition out of Danau Bira (Lake Holmes), just west of the isolated Foja Mountains. Danau Bira was the site of a large mission training camp run by the Summer Institute of Linguistics. I got permission to spend several days at Danau Bira to check it out.

I booked a seat on the weekly Mission Aviation Fellowship flight from Sentani to Danau Bira. In this way I could make an overflight of the Fojas, to look for possible helicopter landing sites in the mountainous interior of this isolated range. We left Sentani airstrip early on a bright and cloudless morning. The pilot took a bearing due west from Sentani, with the thought that we would pass just south of the main peaks of the Fojas.

For the first twenty minutes we flew over undulating hill forest, with a few signs of logging toward the coast. Looking straight ahead, I could see the magnificent dark mountain mass, rising abruptly from the low forested plain. From this vantage point the physical isolation of this range was striking. The high central range and Freeport mine were off to the left, separated by the broad expanse of swampy lowlands of the Mamberamo Basin. To the north was the coast and the Pacific.

Out of deference to my interests, the friendly pilot nudged the plane a bit northward from his normal course and soon we were over the southern scarp of the Fojas. For twenty minutes the pilot flew westward and I gazed down on the uninhabited mountain range, its forested slope broken only by the occasional landslide or river course. There was no sign of human habitation. As we were approaching one of the western high summits (about 7,000 feet), I observed a nearly circular opening atop a rounded ridge east of one of the main summits. Touching the pilot on the shoulder, I pointed toward the clearing; he quickly circled it while I snapped photos. I made notes and quietly celebrated—we had an insertion site. The remainder of the fieldtrip was uneventful but provided useful information for our planning. It looked as if things were moving forward on our ambitious plan.

In 1991 I departed the Smithsonian to work for Conservation International, and at that point our repeatedly delayed plans for a Smithsonian expedition to the Foja Mountains had to be shelved. Conservation International was a young and growing nongovernmental organization (NGO) focused on conserving the biodiversity of the richest and most threatened environments on earth. The good news was that my new boss, Adrian Forsyth, had an interest in organizing a series of Rapid Assessment field trips to Irian Jaya, including to locations such as the Fojas. In preparation, he and I decided to make an Irian reconnaissance in 1995.

We traveled east from Jakarta with our Indonesia director, Dr. Jatna Supriatna, and a prominent botanist from the Indonesia Institute of Science, Dr. Dedy Darnaedi. Our first destination was Jayapura, the capital of Irian. Jayapura, in a picturesque bay on the north coast, nestled against Indonesia's eastern border with Papua New Guinea. After a couple of days there meeting with Forest Department officials and other agency staff, we were scheduled to do a series of reconnaissance flights in a chartered Twin Otter.

We had the plane for two days, to conduct a series of overflights of places of interest for the biodiversity survey. Conservation International routinely follows this procedure when it first enters a new country, to scope out the resources. First we flew over the Foja Mountains. The flight was disappointing because of a combination of heavy clouds and poor communication. We never located the little upland clearing I had seen in 1987. The pilot, banking heavily and circling tightly over the massif, only succeeded in making us dizzy and nauseous. We had better luck flying over the Wapoga Mountains in the afternoon. Wapoga became the site of a Conservation International Rapid Assessment expedition in 1998.

The main flow of the Mamberamo River, draining the vast roadless
Mamberamo Basin

Over a two-day stretch we were able to take close-up looks at the many
wonders of Irian Jaya—one of the last wilderness lands on earth, where un-
trammeled forest lands and waters predominate. The abundance of Irian's
natural features was best seen from an airplane: high craggy peaks, flat expanses of tall jungle, plateaus and escarpments, waterfalls galore, rugged
mountain ravines, hidden valleys, fjordlands, white-sand barrens, weird mud
volcanoes, and more. It was breathtaking and humbling. There was a lot to
survey! It would require decades of work.

For several days we were based in Timika, Freeport's lowland base
for mining the vast mountaintop ore deposits. After our ill-fated 1981 trip,
Freeport had built a five-star Sheraton Hotel on a patch of land carved out
of the Timika jungle. We stayed there, meeting with a range of Freeport staff
and exploring the possibility of working with Freeport on various research,
education, and conservation issues in Irian. It was somehow unsettling to
see this luxury hotel in the New Guinea jungle. The Sheraton backed onto
tall lowland forest, through which the hotel staff had constructed a small

nature trail. I walked this path each morning, delighting in the six species of birds of paradise and many lowland forest birds. I had an abundance of time that had not been available to Brian Finch and me when we first visited the Timika lowlands in 1981. This time we were here under considerably different circumstances. We pored over maps with Freeport geologists and environmental scientists, and discussed what they had learned in their years of exploring this corner of Irian. They were as excited as we were by the biodiversity potential of this province.

Today, in 2008, we realize we are still just scratching the surface of Papua's biodiversity. We have made a lot of progress on the conservation front, which has led to the designation of new conservation areas. In addition, some of the existing protected areas have management plans and resources to ensure their protection. But there is still a long way to go. Getting the job done is the work of a lifetime.

The Freeport mine has attracted more attention, virtually all of it negative, than perhaps any other modern gold mine. Among other things, it has been accused of standing by while the Indonesian military brutalized and imprisoned tribal people in and around the mine. It has also been accused of wreaking environmental havoc on the watersheds that drain the huge open pits of Ertsberg and Grasberg. In the late 1990s the mine was accused of inadvertently creating social dislocation when instituting a beneficial fund to aid development in and around the mine. Additionally, the mine has been blamed for the mysterious murders of several local residents (including an American) in an attack that the military asserted was led by "West Papuan rebels," but which the world came to believe was the doing of the military itself.

What to make of all these accusations? Over the last decade I have met several times with a leader of the Amungme people, Thomas Beanal, who was instigating a lawsuit in a Louisiana court against Freeport for loss of tribal assets as a result of the mine's environmentally destructive practices. In 2001 I met with Goldman Award winner Mama Yosefa, who had been imprisoned for months by the military at Timika because of her lobbying efforts against the mine. I have spoken with a range of stakeholders and, frankly, there is no simple story. Giant mines inevitably have major impacts on local watersheds and on the traditional societies that live near mine sites. Take a place as remote and forbidding and undeveloped as the southern scarp of New Guinea's highest mountain range. Build two cities, an access road, a port, a pipeline, and open pit mines on two mountaintops, and you have the recipe for significant environmental and social impacts. Throw in a meddlesome Indonesian military fixated on the Papuan separatist movement and money-making, and you have the makings of a humanitarian disaster.

Making gold pay in such a remote location is bound to have high costs. The question the world must ask is, do the benefits offset the various costs? The answer varies, depending on which stakeholder is replying. The main stakeholder is, of course, the national government of the Republic of Indonesia. Its viewpoint will likely be quite different from that of the Amungme and Kamoro people who live near the mine.

Freeport mine is therefore a huge, juicy target of advocates for the environment and indigenous peoples. It is like a giant Gulliver, attacked by swarms of Lilliputians. Often called the biggest gold mine on earth, it is Indonesia's single largest earner of foreign currency. It would be a tall order indeed to bring Freeport down. Is it a monstrosity? Of course. Any huge mine is a monstrosity, by its very nature. Mining is a dirty, messy business, but one that produces great wealth for nations.

Freeport has been a particular target because of the bumptious nature of its long-reigning chief executive officer, James Robert (Jim Bob) Moffett, and because of the association between the Freeport mine and the Indonesian military. In the past, Moffett has taken considerable pleasure in playing the bad guy to the environmentalists of the world (probably not particularly wise for Freeport's stockholders). The military connection is more tenuous but also more deadly. Given the nature of the Suharto regime, what choice did Freeport have with regard to the presence of the military? Could Freeport have kept the military at arm's length? Unlikely. No one disputes that mischief makers and rebels have, over the years, attempted to sabotage the Freeport

The Victoria Crowned Pigeon, a heavily hunted
denizen of New Guinea's lowland forests

operation. As an example, the slurry pipe that delivers the ore to port has
been cut on several occasions.

Mines, especially large mines, have a huge impact on the tiny and
little-developed societies that inhabit the forests in which the ore body happens to be located. The usual result is destruction of the way of life of those
hapless individuals who happen to be the landowners. These tiny societies
are swamped by the influx of resources and cash, and by the freeloaders,
carpetbaggers, and opportunists who arrive in the area, drawn by the lure
of gold and jobs. Disease, crime, social decay, military brutality, and other
forces bring about the destruction of a fragile society, a sad process that time
after time the world has largely ignored.

The Kamoro and Amungme must lament the common natural resource
legislation that bequeaths subsoil mineral assets to the mining ministries of
the country in question, not to the people who own the land itself. Such has
been the legacy of Western law, inherited by many of the developing nations
of the world, Indonesia and Papua New Guinea included. Local landowners are in no way masters of their own destiny when it comes to mining and
mineral exploration. It must be painful to learn, as a landowner, that you own
the forests and waters and wildlife, but not the oil and gas and minerals lying

just beneath your feet—and that, by the way, all the forest must be removed to get at the mineral wealth. The mineral laws even trump protected-area laws, and many parks around the world have found that mining companies own the rights to the underground assets, park or no park.

In a decade or two, Freeport's Grasberg lode will be finished. The low-grade ore will be processed, the payable minerals will be exhausted, and the mine will close. That will be the end of the mineral benefits. For the shareholders of Freeport, the company will presumably move its assets and equipment and staff to search out future profits in the form of newly discovered mineral resources. For the government of Indonesia, no more revenues will be produced by those ancient and nonrenewable ore bodies. The key question is whether the royalties generated in the active mine years have been tucked away in the national treasury for a rainy day.

For it is those great mountaintop ore bodies that constitute the *natural capital* that generates the mine's wealth. Once depleted, there is no renewal, though there will be decades of postmine costs: polluted water, lost fisheries, land restoration, and the like. What so many governments of developing nations fail to realize is that the annual payments generated by mines such as Freeport are not really "dividend payments" but capital payouts. To be managed properly as assets of the state, these payouts must be managed as capital, not interest or income. In other words, *all* the mine royalties and payments should be reinvested, and the government should be spending only the interest generated by their investment. Then, and only then would a nonrenewable resource like the Grasberg lode provide sustained and sustainable benefit to the people of Indonesia.

The giant British company BP (formerly British Petroleum) is now investing in Papua's next major natural resource gambit: the Tangguh gas deposit, under the waters of Bintuni Bay, to the west of Freeport by about 175 miles. BP is a forward-looking company seeking to burnish its sustainability image, so it is trying to develop the Tangguh reserve in a way that will be responsible both socially and environmentally.

BP seems to have gotten off to the right start. Still, troubles lie around every bend. Keeping the military out of Tangguh is one ongoing issue. Another is Papuan sovereignty versus central (Jakarta) control over natural assets. BP certainly would like to see the benefits of the liquid natural gas project paid out in ways that will first benefit the local communities, then the communities of Papua (Irian) as a whole, and finally Indonesia as a nation. Tough negotiations may be involved to make this happen, mainly because Jakarta tends to pull the strings. Although it is the moral responsibility of the Indonesian government to ensure that development takes place for the benefit of all of the country's stakeholders, past experience indicates that "best practices" will probably emanate from the international corporation, which is more nimble and more concerned about the complaints of its shareholders should it be caught misbehaving overseas.

Many intellectuals in the West would like to see "West Papua" evolve into a freestanding, independent nation-state, as Papua New Guinea did in 1975. In 1960, the Dutch envisioned a West Papua distinct from Indonesia, but President Sukarno had another idea. He wanted all of the Dutch East Indies to become Indonesia, and it is likely that he understood the vast wealth of western New Guinea's forests, lands, minerals—and open spaces, into which the burgeoning populations of Java and Sumatra could translocate. For thirty years Indonesia's transmigration scheme reduced population pressures in the western islands by shipping the poor eastward. Transmigration settlements were carved out of rainforest blocks in many locations around Irian Jaya. They are still visible today in many of the periurban wastelands around most of Papua's major towns. (Look for the clear-felled squares of land marked with rigid grids of roads and tiny aluminum roofs of the settlers' homes, one after the other—Indonesia's version of a Levittown for the poor.)

Peaceful transition to a politically independent and sovereign West Papua is unlikely anytime soon. Indonesia is still smarting over its loss of East Timor. In the unlikely event that West Papuan separation moves ahead, East Timor is the most likely model of transition—blood and gore and eventual UN intervention. Because West Papua has so little international support (most Americans have never heard of it), probably little will happen; the

growing Indonesian military presence will ensure that it remains an essential
component of the Indonesian economy.

Is this bad news or good? A look nextdoor to Papua New Guinea gives
reason for caution. Papua New Guinea has been declared a "failed state" by
some Australian political economists. PNG still has the potential to save itself,
but only through some extreme measures taken in a timely fashion. At the
time of this writing, there is little evidence these correctives are being taken,
in spite of some Australian intervention grudgingly received by the Papua
New Guinea government.

Papua New Guinea's people are not better off in 2008 than they were
in 1980, five years after independence. The economy is stagnant, jobs are
few, and crime continues to trouble all the urban centers. Papua, by contrast,
seems to be in much better condition in 2008 than in 1980. Can it successfully make the transition to an independent West Papua without risk to its
people's well-being? Or is it short-term risk for possible long-term benefit?
Memories of the destruction that Indonesian militias wrought on fledgling
East Timor gives one pause. It is unlikely that Indonesia would relinquish
Papua without a fight.

Given the highly uncertain long-term benefits that foreign extractive enterprise brings to developing nations, one might ask what rational alternative
exists for governments desperate for "good" development. Perhaps the most
difficult but most sustainable model is the one employed by both the United
States and China: do the development yourself. The keys to sustainability are
the implementation of one's own innovations and the development of one's
own workforce and capital based on local investment, especially sweat equity.
Without their own capital, workforce, and centers of innovation, developing
nations will be forced to be the low-wage production centers for manufac-
turing. The alternative, as with Papua and Papua New Guinea, is to serve
as the resource supermarkets, attracting foreign loggers, miners, and fishers
to harvest the timber and gold and tuna and pay a royalty for extraction of
these resource lodes. The tuna and timber can be managed as renewable

resources, but that is best done by tightly controlled local industries managed internally for the long-term good of the nation. When outside interests drive the process, overharvest and failed stewardship usually result in a degree of depletion that can never be reversed. The lessons of Papua New Guinea's corrupt forest industry should be carefully studied by genuine political leaders in Indonesia and Papua.

CHAPTER FIVE

Biodiversity and Intrigue
across the Inner Line

A flock of tree-pies flew over, and I followed them

into the jungle. In a clearing I found a small party of

rose finches, refugees from the wintry weather at high

altitudes, who would make their home in this valley.

S. Dillon Ripley,

Search for the Spiny Babbler (1952)

INDIA CONSISTS OF A LARGE triangular breakaway section of the ancient supercontinent of Gondwana, now sutured onto Asia by the phenomenon of plate tectonics. The glue holding India to Asia is the mighty Himalayan chain, the zone where proto-India, after breaking from Madagascar and Africa, collided with Asia about 45 million years ago. For biologists and zoogeographers, these areas of contact between ancient land masses are the most fascinating of places—biotically rich, complex, and centers of biological evolution. These areas of vast mountain-building (the Himalayas, the Andes, the New Guinean central mountain chain) are the earth's great biodiversity generators.

In 1988 I was asked to organize one last Ripley-Smithsonian reconnaissance to the subcontinent, this time to India's northeasternmost corner, where the eastern Himalayas penetrate into northern Burma. Called Arunachal Pradesh (Province of the Sunrise), this is the least studied and richest corner of India, home to some one thousand Tigers in the wild and other natural wonders. It was also known as one of the most difficult of access because of Indian paranoia about its borders.

This field trip to India would be my fourth, and the most complex. Ripley had been working with Sálim Ali since 1947 on a comprehensive study of the bird fauna of the Indian region. Their work in Arunachal Pradesh and Bhutan was the culmination of a four-decade collaboration. Ali and Ripley completed their Bhutan studies in the 1970s and commenced studies in Arunachal in the 1980s. These mountainous frontiers constituted the Indian region's ornithological promised land.

Ripley and Ali had already completed two field trips to Arunachal Pradesh. This trip was to be their last and most comprehensive survey, to the very easternmost tip of Indian territory, which lies east of Mandalay and Rangoon. We were to survey the eastern verges of the vast Namdapha National Park, in a narrow peninsula of Indian territory. This easternmost finger of India was surrounded on three sides by Burma. Since northeastern India was off-limits to Western scientists, we were visiting a forbidden land, across India's "Inner Line."

The Inner Line is the Indian bureaucratic term for certain of the country's international frontier areas that are politically sensitive and thus under

Preceding page: Village and mountains, Arunachal Pradesh

tight control. Usually a foreigner cannot cross into an Inner Line area without special permission. We were able to secure permission because Ripley had a special relationship with the Gandhi family. He successfully petitioned Prime Minister Rajiv Gandhi for clearance to conduct zoological surveys in and around Namdapha National Park. Given the rocky relationship between India and the United States in the 1980s, it is probable that Ripley was perhaps the only American with the stature to obtain such permission. Still, authorization directly from the prime minister does not necessarily translate into full support of the decision at the many lower levels of India's byzantine bureaucracy.

Tell any normal person that you are making a field trip to Arunachal Pradesh and you will most likely be met with a blank stare. But tell any naturalist with a knowledge of East Asia that you are going to Arunachal Pradesh and you might see a look of jealousy. Almost any naturalist interested in plants or animals would give a pinky finger for a chance to visit this province, one of the richest and most inaccessible corners of Asia.

Arunachal is special because it marks an immense southward bend in the Himalayas. Several of Asia's rivers are pinched together here in an accordion-like series of parallel ranges caused by the compression of proto-India's northeastward impact into Southeast Asia. Thus, we find the mighty Brahmaputra just to the west of Namdapha, then the Irawaddy, Salween, and Mekong to the east. This concatenation of great Asian rivers and high, rugged, rainfall-drenched ranges, has created Asia's simmering cauldron of evolution and speciation. The area between Bhutan and southwest China's Yunnan and Sichuan provinces is the heartland of the Asian biota; no place on this massive continent is comparable. It is the meeting point of Asia's four quadrants: South Asia, Southeast Asia, Central Asia, and East Asia.

Over the years India has been remarkably paranoid about allowing visitors to the far northeast (the states of Arunachal, Nagaland, Manipur, Tripura, Mizoram) because of China's standing claim to the northern section of this territory, and because of the secessionist activities of local ethnic groups in the region who wish to separate from India. The area is ethnically hyperdiverse, mainly with non-Hindu tribal groups of East Asian stock who feel little solidarity with India. Look at the map. India's far northeast is a big oblong appendage isolated from the rest of India by Bangladesh and Bhutan, connected only by a narrow neck of Indian territory in northern West Bengal state that is probably no more than 20 miles wide—a tenuous linkage indeed.

In 1988, getting there with a large heap of equipment and field supplies was no easy task, in spite of the voiced official support of the standing Indian prime minister, the active participation of India's Zoological Survey (ZSI), and the approval of that highest god of Indian ornithology, Sálim Ali.

Arriving in New Delhi predawn after my long flight from the United States. I passed through customs and security without event. This was my fourth field trip to India, so it was getting to be old hat. The hotel where I had hoped to stay was full, so I settled for the dingy Ashoka Palace in a west-end suburb of Anand Niketan. After a few short hours of sleep I was up and out to the U.S. embassy to check in with the science office staff there. My reunion with S. Subramaniam (Suby) was upbeat and cheerful, but the science officer, George Noroian, had bad news: our four huge cardboard boxes of air-freighted field equipment had been impounded by customs, which demanded that 100,000 rupees duty be paid before our property could be released. This sum amounted to $7,500, a charge that had not been budgeted. What to do?

I spoke with Biswanath Roy, my main contact at the Zoological Survey of India in Calcutta, and was told that all arrangements at his end were proceeding well. S. Saha and K. Mishra, my two field zoologist colleagues from ZSI, headed to Assam that day by train in order to begin advance work for the expedition.

On my second day in India, I did some shopping and returned to the embassy to hear that the science team had made no progress with the intransigent bureaucrats in customs. On my third day, I spent the morning at the embassy, attempting to persuade the science team to make some extraordinary effort on behalf of the expedition. Of course the team had a half dozen other competing issues, so they said they would do what they could. The science officer merely repeated his gloomy opinion of our predicament. I sent half a dozen cables in various directions in search of clarification, assistance, resolution. I began to formulate a new timetable that took this cargo disaster into account. We could not launch the expedition without those four boxes of camping gear and field equipment.

On my fourth day in India, I was up early to fly to Calcutta to focus on final in-country plans with my partners from ZSI. I would return to New Delhi once the snafu with the boxes was resolved. The flight to Dum Dum was delayed three hours. In Calcutta I stayed in a quaint tourist hotel, the Lytton. There B. Roy and I began the detailed planning and pricing of the expedition. Roy would handle much of the financial and logistical work that made the field trip possible. Sleep that night was difficult because Calcutta is a remarkably noisy place, and some sort of holy celebration was taking place that involved thousands of firecrackers.

On my first day of preparation, I met with Roy, Saha, and Mishra to iron out the last details of the travel plan. What became clear was that there were nagging doubts about permissions from the Arunchal Pradesh state government. I might have to visit Itanagar, Arunachal's state capital, to cut some red tape. How many days would we lose to such a venture? I dined that night with a relative of Roy's in one of Calcutta's upscale neighborhoods. The area was quiet and orderly, but the middle class in West Bengal were living a lifestyle quite distinct from what I knew in the United States. Conditions were spartan. Houses were open to the elements, and air-conditioning was nonexistent, in spite of Calcutta's summer heat and humidity. The food was excellent, and so was the hospitality. But I felt I was under a microscope, as Americans of course were rarely houseguests in 1980s West Bengal (a state run by the communists). Later that night I wandered around the tourist section of Calcutta. It was everything I imagined it might be—crowded, smoky, noisy, jammed with cars and trucks. My little hotel was an oasis of relative purity in this center of sordid urban life.

On my sixth day in India, I joined Mishra to cash my rupee check from the PL-480 program. It turned into a several-hour bureaucratic ordeal, thanks to procedures at the Calcutta bank we visited. The afternoon was spent at the Indian Museum, visiting Saha and also meeting Dr. Biswamoy Biswas, another member of India's ornithological pantheon.

On my seventh day, I took the early flight back to New Delhi. It was Republic Day and Delhi was filled with parades and pomp. I watched some of it on television (the real thing was less than a mile from my hotel) and napped. I saved my energy for the next big push. On Day 8 I visited Suby at the embassy to learn that there had been no progress on the boxes, and no word from the United States on payment. What frustration!

On Day 9, joined by the embassy's science officer, I visited the Ministry of Environment and then the Ministry of Home Affairs, in order to obtain paperwork that would allow me to travel into Arunachal Pradesh—the all-important Inner Line permits. The day was marked by many small glasses of milky, sugary tea. On Day 10, I was at the embassy yet again, and yet again there was no good news. Suby visited the Ministry of Home Affairs, and I did some shopping for little things, but still we did not have our boxes of gear, nor did we have my letter of transit.

On Day 11, I was stricken with a nasty Delhi cold and did not leave my hotel room. Day 12 was a Sunday, a free day for which I had planned a field trip to get out of Delhi. I headed to Sultanpur sanctuary, in neighboring Haryana state. The taxi driver was uncertain about the route, and there was no road sign of any sort anywhere in the hinterland. Every two-lane road through the barren agricultural countryside looked alike. We did what one does in any rural spot, repeatedly stop and ask directions.

In spite of the delays we arrived to witness the coming of dawn in the chill late-winter fog. Sarus Cranes were giving quiet bugling notes and moving off to forage in groups of five to ten. At first light I walked through the mists out onto the drying lakebed, watching the silhouettes of flamingos in large numbers stretching and dipping their bills into the water for the first feed of the day. In the cool and sunny morning on this vast open plain, more than two hundred Common Flamingoes and a single lovely crimson-legged Lesser Flamingo fed, surrounded by Avocets foraging in the shallows. A herd of stately Nilgai antelope lingered in some shrubbery. The big blue-tinged, short-horned bull males were massive. They had an African look, a reminder of how much the Indian biota shares with Africa (the flamingos, but also much more—Lions, Leopards, other antelopes, hyenas, jackals). Sultanpur Lake has a magnetic attraction for ducks, geese, waders, partridge, and various songbirds; it has standing water in a vast plain where water is scarce.

Day 13 was a living hell in pursuit of customs clearance for the four giant boxes. Four of us waited five hours at the customs office, only to be told to return the next day. We had spent the day petitioning an array of petty Indian bureaucrats. The final clearance, so close at hand, was denied us at 5:15 P.M. because the second-highest customs official in all of India had decided to leave the office early.

Day 14 found us at customs early, chipping away at the impounded airfreight issue. By one o'clock we had our paperwork, having paid the slightly reduced duty of 66,000 rupees. We rushed to the *godown* (storage shed) to pick up our boxes. Repacking them right there in the parking lot, we quickly took them to the airport to send them on their way to Calcutta. I followed in the evening, finally on the first leg of my journey to Namdapha. Two weeks of frustration were over, and I slept peacefully at the airport hotel in Calcutta.

I was up at 5:40 A.M. for the flight to Assam, gateway to Arunachal Pradesh. How wonderful to be free of New Delhi and its horrid bureaucrats! After an unscheduled touchdown in Tezpur (Assam's capital), we winged onward to Mohanbari, the airport for Dibrugarh, eastern Assam's main commercial center. We dropped our bags at a government rest house, and Roy and I headed to the superintendent of police to present my credentials. We remained in Assam, but our rest house was owned by the government of Arunachal Pradesh. We met some Arunachal officials who were quite friendly, and everything seemed set for our military charter in to Vijaynagar in two days' time. Finally, things seemed to be in order.

I spent the next day repacking for the charter, while my colleagues from ZSI tidied up loose ends related to the field trip. It was a warm, sunny day, and when we had evening drinks with Tape Bagra, the assistant district officer for Namdapha, I could relax and enjoy myself, because things were finally going our way.

On flight day I was up at dawn, wakened by the tea boy, who brought hot tea to my room. The day broke cloudless. We arrived at the military airfield at 8 A.M. and waited two hours to board the big Russian transport. We false-landed at this airfield three times to burn fuel (so we could land on Vijay-nagar's short field). We then made the short flight over Namdapha Park, with views of a jumble of forested hills following the course of the Noa Dihing River. The huge mass of Dapha Bum peak (15,000 feet), capped in a mantle of snow, rose majestically just to the north of the river. Except for the snow, the thick green forests looked tropical, much like what I had come to know in New Guinea. We circled the Gandhigram Valley three times before jolting to a landing on the gravel-and-dirt runway. Piling out of the plane, we dumped our gear and food at the side of the landing ground. A swarm of people came to carry our gear to the government rest house beside the airfield. From its little porch I could look across the flat agricultural valley to low forested hills and upward to snowy peaks to the north, along the border with Burma. I was awestruck to realize I was standing in the geographic center of the Asian realm, at one of the most isolated and rarely visited frontiers on earth.

Aside from the subtropical humid forest on the nearby hills, there was little here to remind me of New Guinea or Panama or even India. Although there were a few Indian government bureaucrats in the tiny hamlet surround-ing our rest house, most of the faces I saw looked Tibetan or Mongolian. In fact, they were Lisu tribal people, who as members of a Christian mission sect had migrated from China and northern Burma to colonize this little Shangri-la in India. They had found a little finger of Indian territory, outlined by a perimeter of sharp wooded hills that jutted into the eastern verges of northernmost Burma, not far from the town of Putao, famous historically for its ruby mines. This area was the northeastern extension of the Tirap Frontier Division, just south of the Mishmi Hills, which adjoin southeastern Tibet. The eastern headwaters of the massive Brahmaputra drainage received huge amounts of rain—probably in excess of 12 feet annually.

The humid climate forest, the mix of hills and rivers, and the great ele-

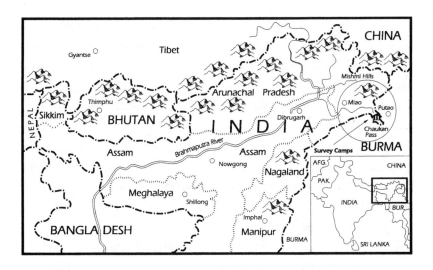

vational range has produced a zone rich in plants and animals. We were here to survey the eastern boundary of Namdapha National Park, one of India's network of Tiger reserves, with at least sixty Tigers recorded on the most recent survey. Ali and Ripley's earlier trips had taken them to the western side of this mountain-clad reserve. Our preparatory research had indicated that this eastern area should be rich indeed in birds and mammals. We had visions of a tropical paradise with six species of cats and perhaps six or more species of primates. Our guess was that we were the first naturalists to conduct bird and mammal surveys in this outmost frontier of northeastern India. After my close encounters in Andhra Pradesh, I was particularly interested in seeing a Tiger or some of the other native cats in the wild.

Presumably prior to the arrival of the Lisu people, this valley was entirely forested and the stomping ground of elephants, Tigers, Leopards, various deer, and primates, but no resident humans. Certainly during World War II troops must have walked through the valley on patrol. We were about 25 miles north of where the famous wartime Ledo Road cut from Assam into Burma, built under the guidance of the iconoclastic American general "Vinegar Joe" Stilwell. During that war many Allied supply flights went from India to China "Over the Hump," and this was the hump they flew over to get there. In the 1960s the Lisu people arrived and cleared the valley floor for rice and vegetable crops. They made the valley habitable. Without their clearings and the airstrip, it would have taken us two weeks of trekking from Miao (on the west side of Namdapha) to get here, and we would have needed a whole train of elephants. This was a part of the world where elephants were still used for jungle transport.

I took a hike down valley to the border of the park (about 7 miles) to get my bearings. It was wonderful to be stretching my legs in the open country, far from any crowded Indian city. I flushed a fireback pheasant foraging along the trailside, the first of what would be scores of encounters with bird species known to me only from illustrations in books. At the eastern park boundary, the bordering hills pinched inward and the Noa Dihing tumbled into rugged forest country. Isolated from the outside world, walled in on all sides by mountain forest, we were in a little enclave—our research laboratory for the next five weeks.

The Ripleys and mammalogist Don Wilson were due to join us shortly. They would have the luxury of an established field camp, which we would be constructing over the next few days, once we agreed on a spot. I quickly discovered in conversations with the local officials that the Lisu people were prodigious hunters, using mainly homemade crossbows that fire little bamboo darts, which they tipped with a locally manufactured poison. The news disheartened us, because it meant that even in this isolated spot wildlife had already been heavily cropped. Our survey work would be more difficult than we had realized.

After a very long day of looking, we chose a spot for a campsite. By that time it was late in the day, and we had to hike back to the guest house in the dark and cold, stumbling, grumbling, and hungry. The hot meal that night tasted particularly good, but we still were chafing to get to work.

We set our first field camp at Ramnagar, halfway between the airstrip at Vijaynagar and the Lisu town of Gandhigram downstream. The area was sparsely settled, with few houses, although most of the bottomland forest had been cleared for rice and other staples. We set up aside the Noa Dihing River in the far corner of a dry rice field, with a nice view of the snowcapped

peaks to the north. The camp consisted of three very large nylon-wall tents
that we had customized to withstand the high rainfall. In addition, we had a
screen tent to serve as a work center (with tables and chairs). And Roy had
constructed a mess area with a large plastic roof, where we took our meals.
We had two smaller tents for the field assistants. Finally, we constructed a
pit toilet in a small patch of scrub. This was a field camp fit for the olden
days—capacious and well provisioned. We had ten support staff, whose job
it was to keep the camp running efficiently: a chief cook, several assistant
cooks, washers, porters, guides, and local naturalists.

We now had five weeks to characterize the bird and mammal life of this little
valley. In reality, I would have needed a year or more to obtain a detailed
knowledge of the complex avifauna, but we had to make do with the time we
had. We would work seven days a week, from before dawn until after dusk.

My job was to lead the bird survey work with S. S. Saha, assisted by
Mishra and some Lisu field informants. It was more art than science, using
any and every means possible to encounter and identify each bird species in
the valley at that time. Mist-nets and ad lib field observation constituted 99
percent of our work. We opened the nets at dawn, then walked the trails looking for bird flocks and solitary species; we returned to the nets to check them
every ninety minutes throughout the day; we processed the netted birds and
continued the observations until dusk. Birding and netting to the maximum
required that we learn the local trails and the little places in the forest and
scrub where birds hid out. When, on the first day, one of our workers came
in with two bamboo partridges he had shot with his crossbow, we knew that
any of the wild creatures here were "fair game" and probably much over-
harvested. Only the most elusive and shy species would be left in numbers.

Nonetheless, every day in this valley I encountered some bird or mammal that I had never before seen. This wonderful new Asian environment was
oddly familiar because of the presence of maples and oaks and other temperate-
zone trees. What more could a naturalist want? In fact, the first few days were
dizzying with the confusion of new wildlife. I took lots of field notes on birds

Tent camp near Ramnagar, alongside the Noa Dihing River of Arunachal Pradesh, with Burma border ranges in the distance

I encountered in the day and then made identifications from my notes back in camp after dark. Most species, even the ones new to me, were at least a little familiar, because I had spent a lot of time in New Delhi studying Ali and Ripley's *Pictorial Guide* to Indian birds. Their book helped a lot, but invariably there were puzzles. Some of the species groups in northeast India posed tough identification problems, in particular the grass warblers and the *Phylloscopus* tree warblers. But nothing was more exciting than trying to crack the code on an unidentified bird species. One always hoped that it might turn out to be a species or subspecies new to science.

One day a large transport plane, followed by a huge military helicopter, passed over our Ramnagar Camp toward the Vijaynagar landing ground 5 miles up the valley, carrying as passengers the remainder of our team: the Ripleys, Don Wilson, and additional Indian scientific counterparts. We expected the group to arrive in a few hours, but there was no sign of them as dark closed in. The next morning we learned that the Ripleys had canceled because of an injury Dr. Ripley suffered just before leaving Washington. Wilson arrived with four Indian colleagues, Dr. Agarwal, Bharat Bhushan, Mr. Chandiramani, and Mr. Britto. We were finally at capacity. After five days of netting we had recorded 117 species of birds and netted 200 individuals. On this particular morning, the Hoolock Gibbons gave a songfest in the hill forest just behind the camp. These great apes were wonderful vocalists, producing musical whooping notes in a cacophonous family group. It was electrifying to hear and excited all the naturalists in the group, especially Wilson. Mammals had been scarce so far.

Each new day promised new birds, and I was never disappointed. Checking the mist-nets always produced surprises. It was impossible to predict what we might find. Some birds—such as the Nepal Babbler, Black-chinned Yuhina, Golden-headed Babbler—we captured over and over (we could tell they were retraps, because we clipped their outer tail feathers). Others we encountered only once or twice. The first few days that the nets were set in a given location they produced a lot, then the take tapered off, forcing us to move the nets every several days. Our Lisu team members quickly became proficient at this task once they saw how it was done.

Everybody had a job, and people were busy every day doing their specific task. We went our separate ways and usually saw one another only at meals. The evening meal was the most crowded, because almost everyone would have returned to camp and bathed before the food was served. Roy made certain this complex field camp ran well. We had our own little town on the outskirts of tiny Ramnagar, our natural history factory, working at full production. Our camp was the most fundamental expression of field natural history, the primary inventory of a never-studied site. We were following in the footsteps of Whistler, Blanford, Hodgson, Hume, Blyth, Jerdon, Horsfield, and Baker, the colonial founders of Indian zoology.

The birds proved to be the source of greatest interest. Over the thirty-seven days we spent in the field, at four different camps ranging from 3,300 to 8,250 feet above sea level, we recorded 236 species. The forests and shrubbery and streamsides of this hidden valley were infested with birds: bulbuls, bab-blers, parrotbills, partridges, woodpeckers, kingfishers, flycatchers, robins, tits, leaf warblers, mynas, and forktails. It was a paradise for colorful elusive and obscure landbirds familiar only to aficionados of birds of the Indian subcontinent.

Several birds were first-records for the state of Arunachal Pradesh (Indian Jungle Nightjar, Orange-billed Jungle Myna, Streaked Long-tailed Wren-Babbler, Brown-headed Tit-Babbler, Blue-fronted Robin, Chestnut Bunting); one was a first record for India and a new subspecies (Long-tailed Spotted Wren-Babbler); and one poorly known form we elevated to the sta-tus of full species (Brown-headed Tit-Babbler) because we found it living together with its supposed parent form (Gray-headed Tit-Babbler). Per-haps most interesting were our various encounters with three of northeast India's most peculiar and little-known songbirds, the Wedge-billed Wren, the Slender-billed Scimitar-Babbler, and the Long-billed Wren-Babbler. Previ-ously, these three had been only encountered in life a few times by Western naturalists. The Wedge-billed Wren had last been encountered in 1905, one of East Asia's least-known bird species. Finding healthy populations of these mystery birds was heartening to our conservation-minded team.

We also found that the valley's winter bird-flocks were probably the largest and most species rich on earth. I had never seen such profuse bird parties, of such great size, and with so many species in train. The flocks were dominated by babblers, the most abundant of which might include fifty in-dividuals of one species. These flocks dwarfed those I knew from southern India or in New Guinea or even Panama. Flocking seemed to be the main organizing principle of the eastern Indian hill forest's winter landscape.

A less propitious discovery was strong evidence that human subsis-tence hunting in the valley by the Lisu tribe (numbering only a few hundred) had wiped out much of the mammalian megafauna in three short decades. Tigers and Leopards were gone from the valley (though we were told a Tiger still lurked in the forest a few miles farther east, within the park) and elephants now avoided the area. Deer were scarce, and primates almost im-possible to see (the gibbons were noisy, but exceedingly wary and few in

The Valley of Gandhigram, heavily cultivated and overhunted by its
Lisu inhabitants, with Namdapha hills in the background

number). I had two nocturnal encounters with the ocelot-like Leopard Cat.
In each case, with a spotlight I was able to walk right up to the adorable
spotted wild cat—not much larger than a common house cat. In the second
instance, I stood within 10 feet of the little cat as it sat, motionless, transfixed
by my bright light. I have rarely seen a creature as beautiful as the Leopard
Cat, looking much like a tiny and harmless version of a Leopard.

Overall, the mammals were a real disappointment to our mammal crew.
Whereas the cold weather that predominated seemed to crowd the species of
birds into the valley bottom, enriching our bird surveys, the cold may have
sent the mammals downstream to warmer climes (at least it did the bats). But
it was probably the hunting Lisus that ruined our mammalogical hopes.

In reality, during the time we were in the valley we could only get
a first approximation of the rich rainforest fauna that lived there. At the
time I declared that I would like to spend twelve months monitoring bird
populations in this isolated valley. I still would love to do so. Watching the

seasons change in this Indian Shangri-la would be remarkable, what with the strong elevational movement of the locally breeding species, added to the north-south movement of the Palearctic migrants. For instance, we watched skeins of Bar-headed Geese pass over the valley from Tibet, and all sorts of nonbreeding migrants from the north were wintering here. I suspect the valley would host a remarkable passing of the seasons.

❧

The intrusion of American field scientists across India's Inner Line frontier with Burma was apparently unprecedented, and apparently the source of considerable consternation to local bureaucrats from Calcutta to Itanagar to Dibrugarh. While Don and I blithely went about our business in the field, all sorts of plots were simmering. The bureaucrats never really believed our interest was only in birds and mammals. They were sure we were from the CIA, and they were going to make certain we did not steal the strategic secrets of eastern India on their watch.

One of our Indian colleagues was with the Indian CID (the Central Intelligence Directorate). He was sent to spy on us and report back to his headquarters on all the nefarious things we were doing. He monitored our activities for a few days, until he saw for himself that we were interested only in birds and mammals. At that point he decided to make himself useful, so he joined in our work and was appreciated for his substantial contribution. I don't believe he was happy to be stuck with us in this little Shangri-la, but he made the best of the situation. He was certainly able to report back to his superiors that no secrets had been pilfered.

Another of our team, reputed to be an esteemed local field biologist, proved a more difficult case. He turned out to be one of those self-centered, frustrated people who find no pleasure in the joy of others. It became clear after a short while that, in spite of his high local rank, he had no predilection for field research, very little knowledge of wildlife and nature, and no interest in helping our cause. He was a thorn in our side. We generally ignored him, but at some peril. He brooded. He was jealous of the intense pleasure we got from the work we were doing. And when we shouted about finding a "rare

bird," he took us at our word. Unbeknownst to us, he was regularly reporting back to his headquarters in Itanagar that our team was only interested in
stealing India's intellectual property and decimating its endangered wildlife.
He apparently continued to elaborate on this story week by week, creating
music to the ears of the left-leaning bureaucrats of eastern India. They must
have been rubbing their hands with glee at the thought of catching some
American spies while we plied our dastardly craft in the Indian borderlands.
We were later told that this gentleman was radioing back to Itanagar on a
daily basis with the "dirt" on us.

In retrospect we could understand why. Here was an ambitious Indian
civil servant, who probably had dreamed of a power posting in Calcutta,
Mumbai, or Delhi. Instead, he was sent to Itanagar—India's version of Si-
beria. One can only imagine the bitterness that might well up, seeking some
form of release. Documenting our wrongdoing became this pseudonatural-
ist's passion. He had little or no interest in the natural world and, with his
suspicious nature, it took little for him to convince himself that he had an
opportunity to uncover a "crime" and thereby gain fame as a sleuth.

We Americans learned that something was amiss with only about a
week remaining in the field trip. We were told by our colleague-nemesis that
all of our cameras and film were to be confiscated prior to breaking camp.
Don and I had fits, but Roy calmed us down and reminded us that there
was no point in resisting—better to cooperate and look for clemency later.
We dutifully turned in our cameras and film, all boxed for inspection by the
authorities (though we were never sure exactly which authorities). Then one
of our team muttered that we had a spy among us, and that we were under
suspicion and being investigated. We were shifted from our field camp back
to the airstrip at Vijaynagar and instructed that all our fieldwork had to
cease. I was allowed to go bird-watching, but could not wander far from the
settlement. We were able to set a few nets behind our lodging, but that was
it. What a black cloud settled over our little team!

It got worse. We were essentially packed up, ready to return to civilization, but we could not get a transport plane to come and pick us up. We
were left to stew in our bitter juices.

On the first day of our incarceration, I got permission to traipse about a nearby patch of forest in search of wildlife. I saw little wildlife, but I did encounter a remarkable sight that reminded me of what this place had been not long ago. Midway on our hike, my local colleague pointed out a traditional elephant highway. The twice-annual migratory trekking had cut an 8-foot-wide path through the clay hills. In some spots the dirt banks were 6 feet high on either side of the elephant road; the elephants' footwork had been equal to the work of a bulldozer. It was apparent that the elephant population here not long before had migrated back and forth, up and down this narrow valley, presumably going up through the Chaukan Pass into Burma, and then coming back in season. The appearance of the elephant-hunting Lisus had led to the destruction of many of these beasts, and presumably the remnant herd had taken to migrating along some safer route. We neither saw nor heard an elephant during our stay in the valley, though I had seen elephant dung at the edge of Namdapha National Park.

On Day 2 of our incarceration, I did some prebreakfast bird-watching in the stream valley behind our guest house and was lucky enough to have some superb looks at a Slaty-bellied Ground-warbler in full song—the first time the song of this species had been transcribed. The rest of the day I cleaned and packed supplies in readiness for the arrival of the transport. Things were tense. On Day 3, we were assured that today was the day our sortie would come in to fetch us. The weather was sunny and mild, yet no aircraft came. Clouds gathered in the afternoon, and light rain began to fall.

Day 4 brought heavy clouds brooding overhead, and rain fell all day. I slipped out into the nearby forest in search of wildlife and in a little mixed flock of sixteen species I sighted my first Sultan Tit—patterned in navy blue and deep yellow, with a saucy recurved crest. I took afternoon cocktails with the commander of the military post that guarded this lost world. He was a gracious host, and our conversation ranged widely. Interested in the world around him, this gentleman must have been living in quiet desperation as he waited for his next posting, to a place I hoped would be more hospitable than this one (no cable, no satellite, no VCR, no TV). He showed little concern or suspicion about our presence here. Still no plane!

On Monday, Day 5, cloud and rain blotted out the sky and any hope of a plane. We played ping-pong, bridge, then hearts, until after midnight. On Day 6 we remained socked in with mist, cloud, and rain. It cleared in the afternoon, and I went out in search of birds, adding a little chestnut-headed warbler to my list. On Day 7 the morning broke fair, and I headed into the woods for birds. A military plane arrived and took out Bharat Bhushan and our provincial nemesis. We were not allowed on this plane because it was military-only and could not carry foreigners. There had not been a civilian sortie for a month. It was a wonder this community could survive; these transports are the only means of access other than a hike of 85 miles to Miao.

Day 8 was our day. The weather was fair. I took a short walk along the creek behind our house, but did not stray far for fear of missing our sortie. The plane came in at 11:30 A.M. and we boarded with high anticipation. The pilot attempted to start the plane and nothing happened. Blown fuse! We awaited the arrival of another plane with an engineer. In remarkably short order the second plane arrived (extraordinary, in that we had waited seven days for the first plane). The engineer replaced the fuse in a few minutes and we were off, back toward Mohanbari and civilization as it is in Assam.

On the flight back we had even more striking views of the natural destruction the 1950 earthquake had wrought on the forested hill country of eastern Arunachal. The entire hilly area looked as though it had been shaken violently (which it had); the land and forest were a jumble, almost forty years after the fact. Scientists tend to underestimate the importance of natural disturbances in tropical forest regions. We often label forests where we work as "virgin" or "pristine" because we see no human disturbance. Frost, drought, fire, land movement, heavy wind—only one of these needs to strike a patch of forest every century to have profound effects on the local environment. When we add in the many impacts from traditional societies—slash-and-burn gardening, with the introduction of a range of crop trees that persist after the forest regenerates; human-caused fire; human-introduced animals of various sorts—change, rather than stasis, is the norm.

Witness the situation in the Vijaynagar Valley. We saw not a single elephant, but presumably over the centuries elephants had lived in and migrated through this valley, affecting many aspects of forest structure. In a short while the Lisu hunters wiped out the elephants, but certainly the elephant impact lingered long after the animals themselves were gone. The environment responds to these major changes over decades or centuries, not years. Most environments are nonequilibrial, shifting from one temporary state to the next, each new state unlike anything in the past.

After our flight back to the comforts of Assam, we spent the afternoon cleaning up and organizing in Dibrugarh, at the Circuit House. Early the next morning we had to return to Arunachal to wrap up our expedition officially and sign out with the local officials in charge. We drove for several hours, crossing the Inner Line checkpoint by car for the first time, and stopping to have a look at the sign that marked the beginning of the wartime Ledo Road. We finally arrived in the little town of Miao, at the western entrance to Namdapha National Park and back in Arunachal Pradesh. We were in Miao to meet with the forest warden in charge of this sector, to present our credentials, and to "sign out" from this restricted area.

This clearance meeting was fraught with issues. Hassled about our permits and our cameras and our film, we spent five hours arguing our case. At the last minute our cameras were returned, but the films had been confiscated. Thank heaven I had handed in only unexposed films and had secreted away my precious exposed rolls.

Saturday morning I rose early, had one of those perfect Indian breakfast omelettes, and went out searching for birds in the forest next to our rest house in Miao. In spite of the rain I encountered a family group of Brown Hornbills at a nest, a species we had not seen in our surveys on the eastern side. I observed several individuals bringing food to the nest hole—an instance of communal provisioning of the nestlings. After adding nineteen low-elevation species to our list for the Noa Dihing watershed, we headed off to the airstrip to return to Calcutta and New Delhi.

We thought our trip was finished, but trouble had a way of following us. An article appeared in a Calcutta newspaper questioning our intentions and decrying the fact that the national government allowed Americans into this sacred border zone. In addition, all sorts of trouble arose relative to the museum specimens we had collected. These issues continued to dog us back to Washington, and political matters related to this scientific expedition were not cleared up for many months.

Despite the problems, the wonders of Namdapha never fail to stir my imagination. In 1990 I returned to spend a month in Calcutta visiting the Indian Museum managed by the Zoological Survey of India, where our colleagues Saha, Mishra, Roy, and Agarwal were based. There I studied all the collections made in various parts of Namdapha over nearly a decade as part of a government survey of the bird life of the region. It was a rare privilege to conduct this museum review of a bird fauna as rich as any in Asia. I would love to return to that little hidden valley for a twelve-month stint and count birds every day. Now *that* would be a remarkable year.

What did I learn from this expedition to the leech-infested rainforest on India's borderlands?

First, I learned that the American worldview of science and discovery is at variance with that in India and other developing nations. My view, which I think is typical in the United States, is that discoveries and pathbreaking research can be pursued by anyone who can get the job done. In that vein, Americans for the last seventy-five years have been exploring overseas, and discovering and uncovering—in the kind of entertaining pursuits often featured in the *National Geographic* magazine. This research plays well in the United States, but perhaps not in the nations where the discoveries are made.

The issue is one of scientific imperialism, of well-funded scientists and explorers who live in rich countries scouring the world for new species and cool discoveries, which some say should "belong" to those countries. Advocates in the developing nations prefer to "save" these discoveries until such time as local scientists have the resources and capacity to do the research themselves. I believe this argument has some merit. But I also think of science as a worldwide endeavor, in which international collaboration is valuable. Also, a time element is involved: many of these precious places are under immediate threat. Only by surveying them and making clear to decision-makers that these areas are important can they be conserved before they are inadvertently decimated.

Cultural sensitivities need to be recognized, and collaboration must be real and empowering. It is all too easy to "buy" collaboration and to create a cooperative veneer where no true collaboration exists. This false collaboration is destructive. It is safe to say that in our case "collaboration" with our provincial nemesis was nonexistent and might partially explain his jealousy. Real and lasting collaborations are produced by mutual understanding, mutual interest, and respect—which take work and time and patience. I would do things differently if I were to return to Arunachal today.

Worldviews differ on the issue of biodiversity. The American scientific view is that biodiversity is a global resource that belongs to everyone, and that it is the obligation of each country on earth to work to conserve representative samples of this global biodiversity for the benefit of the earth and the well-being of humankind. Many other nations, however, see their own wildlife and resources as *their* material and intellectual property, to be controlled and held close. The fact that India had that attitude in the late 1980s is certainly one of the reasons our American-led party had trouble with various elements of the Indian establishment.

Perhaps the best way forward lies somewhere in the middle. Nations should be able to benefit from their natural patrimony, but the world should be able to encourage each nation to protect this patrimony for the good of all. The challenge lies in finding the balance of interests.

Consider the international plight of the Tiger, an emblematic species in India if ever there was one. Tigers today inhabit only a few countries in South, East, and Southeast Asia. The entire world is vocally committed to Tiger conservation. All the world knows and loves the Tiger in the abstract, though few conservationists have seen one firsthand, prowling a jungle path. I do not have to see a Tiger in the wild to love the idea of a wild Tiger existing in a tropical jungle. The Tiger, for better or worse, has become a universally recognized symbol. Around the world it has been transformed into a plush-toy creature that haunts the musings and dreams of small children (as it did for me as a child).

Yet the Tiger is the earth's supreme living terrestrial carnivore, one of the few wild species that still stalk, attack, and consume humans on a regular basis. A local Indian village thus will petition the government to kill a man-eater. By contrast, a Swedish tourist might happily invest $6,000 to travel to India to see and photograph Tigers at one of India's famous Tiger reserves such as Namdapha or Kanha or Ranthanbore. Then again, an Indian poacher sees a Tiger in terms of cash value (pelt = $2,500? dried organs and bones = $6,000? claws = $250?).

To ensure the future of the Tiger in the wild, is it possible to balance these competing interests? That is the challenge for environmentalists and conservationists and wildlife agency officials. It is said there are more captive tigers living in Texas than there are wild Tigers in India. To save the Tiger's last wild populations will require close and productive *collaboration* between the wealthy (Tiger-loving) nations of the world and the few remaining nations on earth that support wild Tiger populations. Because of different world-views and cultures, this task will be daunting indeed. It is not as simple as donating to the Save the Tiger Fund or arresting a team of poachers. It will require changing the outlook and behavior of all the stakeholders inhabiting those areas where Tigers still roam, and finding a replacement therapeutic for that population in China which feels the need to purchase Tiger parts under its regime of traditional medicine. The most daunting challenge is to transfer the abundant political will in the West to the Eastern nations, where a myriad of troubles keep those societies from investing sufficiently and focusing political capital on a workable solution to the Tiger's survival. Surely we all agree that the world will be a poorer place when wild Tigers no longer stalk the jungles of Asia.

CHAPTER SIX

Forest Gardens

Off the stern a hornbill was crossing the river,

its flight undulating, buoyant—

four deep flaps followed by a glide—

its plumage black in front and white behind.

Redmond O'Hanlon,

No Mercy: A Journey to the Heart of the Congo (1997)

IN 1999 I WAS WORKING AT Counterpart International, in downtown Washington. Counterpart evolved out of a Pacific network of NGOs and focused on improving the livelihood of communities, especially rural communities. The environment division in this nongovernmental organization was implementing three experimental programs: an urban environmental remediation initiative in the former Soviet Union; an experimental agroforestry system called *Forest Gardens* that was based on the analog forestry system developed in Sri Lanka by Ranil Senanayake; and a coral reef restoration methodology called Coral Gardens, a brainchild of Austin Bowden-Kerby. Because of the substantial prior work of Dr. Senanayake, the *Forest Gardens* program was the best-designed and most promising of these initiatives, with the broadest application through the tropical world.

The objective of our *Forest Gardens* system was to encourage tropical rural farmers to conserve standing forest by focusing their cash cropping on multispecies polyculture, with a permanent overstory of mainly native tree species. In some respects the system was a generic form of shade coffee. Coffee could be (and was) one of the typical understory cash crops with *Forest Gardens,* but the farmer did not focus solely on coffee. Instead he cultivated a suite of understory, midstory, and overstory plant species, each of which produced something of value—a crop, a soil nutrient, improved water retention, or shade. The most devoted advocates of *Forest Gardens* believed this system could entirely replace monoculture cash cropping, but most of us thought it more likely that *Forest Gardens* could be employed to diversify land use and to reduce the predominance of monoculture systems and the rate of local deforestation. *Forest Gardens* plots could be interspersed among the monoculture plantings and placed in the most environmentally vulnerable sites.

Counterpart's senior leadership asked me to travel to Ivory Coast (more properly Côte d'Ivoire) in the heart of West Africa to scope out opportunities to implement *Forest Gardens* there. While working for Counterpart, I had glimpsed East Africa (spectacular wildlife and open plains and grand vistas), savored Madagascar (wonderful birds and lemurs), been surprised by the wealth and modernity and complexity of South Africa (elaborate expressways, tree-lined rich residential neighborhoods, wonderful birding,

Opposite: Piping Hornbills in flight

troubled townships). So now it was on to West Africa and the mysteries of
Ivory Coast.

Trying to picture that country, I could conjure up very little indeed. I
could close my eyes yet visualize virtually nothing except the tourist photos
I had seen of the impressive city of Abidjan, with its coastal lagoons and
skyscrapers. Yes, I knew there were rainforest parks and even game reserves,
but I could not imagine what they were like. The snapshots a colleague had
shown me depicted tropical countryside with tall grass and a scattering of
trees. Those photos could have been of roadside secondary habitat in just
about any equatorial nation.

Because of the strong French colonial connection to a number of West
African nations, my flight was routed through Paris, with a few hours to kill
in smoky Charles de Gaulle Airport. I found lots of elegant perfumeries but
not much choice of food to eat, and a money-changing booth that was closed
the entire morning. The airport was rather new in 1999 but not particularly
user friendly. I always feel handicapped by my lack of French. Deep inside
I believe I should be able to speak the language fluently (I "studied" it early
in grade school). In reality, I can get by with English and waving hands and
my smidgen of French. I keep promising myself I will learn, until I hear the
flight announcement in rapidfire French and realize I will never be able to
decode that slur of vowel-heavy and s-deficient syllables that is so totally
unlike English.

Flying southward across the Mediterranean I napped, while far below the sea
sparkled under a cloudless afternoon sky. Then we approached the north coast
of Africa—Algeria. It was a rugged coastline with rather unfriendly-looking
mountains looming behind the shore, some ridges reaching all the way down to
the water. The mountains were blackish on their heights, where conifer forest
clung to the slopes; elsewhere the land was parched and scuffed by the hand
of humankind—little whitish roads traced through the sharp hills, and small
hamlets tucked into curving beaches backed by the imposing mountains.

Within a few minutes we were passing into lowlands behind the coastal range and then some more interior mountains—the Atlas, I supposed. These were more forested and formed bulwarks trending more or less east-west. South of them lay rugged sun-parched waste, where in a low spot a huge salt-bleached pan spread in no particular direction. Beside this I could see little oases, with water and some sign of life. Then another small range of dry mountains, this time dark brown but apparently without forest. We were now deep into the rugged desert lands of North Africa, passing over land that reminded me of the American Southwest, with some dry-land agriculture, small roads, and even a river with a slight flow of water, passing through a narrow, rocky valley. Deserts from the air display so many of the earth's features and forms and geological phenomena—Freshman Geology writ large from 33,000 feet. Having grown up in the boring coastal plain of the eastern United States, I always appreciated a look at real geology where the earth was evolving rapidly.

Next, I was seeing dry riverbeds that coursed across the landscape, paler than the surrounding surface, with small, wetter tributaries marked by the presence of green vegetation—bushes growing in the streambeds. Everything was baked under the sun's fierce glare, and every unvegetated spot was a dull bleached tan of some hue. Here in the wastes of central Algeria I could see a straight cut through the landscape that was apparently a roadway or pipeline.

With each few minutes that passed we seemed to move into an ever dryer and flatter landscape. The desert became more rust-colored and small puffy clouds obscured my view. I dozed restlessly and awoke to find savanna forest below—small trees, spaced neatly but in profusion. I was somewhere northwest of Ouagadougou, in Burkina Faso, a country once known as Upper Volta. The lesser nations of West Africa are many and little known. Benin, Togo, Burkina Faso, Guinea, Mali, The Gambia, Upper Guinea, Senegal are difficult to pick out on a sketchmap. Better known are Liberia, Ghana, and Ivory Coast.

As we crept southward ever closer to the equator, the habitat gradually greened again. First there was some closed woodland, with little sign of human occupation. Then I saw villages scattered among degraded forest

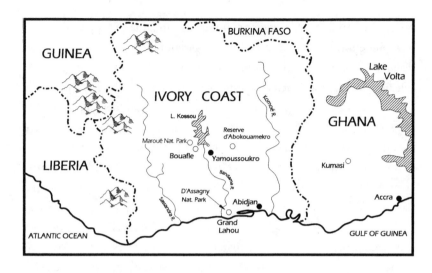

patches and broad avenues of garden land linking the villages. These villages
were much larger than anything I had seen in New Guinea, and appeared
more modern (tin roofs, Western construction). The closer we got to the
coast, the more forest I saw, until we passed over a nice expanse of original
forest (a park?). Finally, we were down by the coast west of Abidjan and
the plane began its final descent. The forest was broken in many places by
villages, cacao, and the large blocks of oil palm, edged by unkempt patchworks of local gardens. Still, some tropical forest remained, especially along
the waterways. As we headed into our approach, the sky began to darken;
I could see white egrets set against blue-black storm clouds, winging over
waterways toward their roost.

It is advisable to arrive in a tropical city in the evening. The sun is setting and
the air is cooler. In addition, the crowds begin to lighten. Making my way
through the airport at Abidjan, I chanced to engage a pleasant and helpful
local facilitator. We pressed through the departure corridor to a dilapidated
deep orange taxi (all the taxis in Abidjan are deep orange), that sped us to
town. There we came up against the last of the evening rush hour. I had asked

my airport helper to join me in the cab and serve as a guide and interpreter. We slowly made our way up the hill to Plateau, the district where I had been told to look for a suitable hotel.

In the taxi I remembered what is special about tropical cities at night. It is the fecund smells emanating from the fields and swamps and homes that ring these metropolises. Manure and burning garbage and swamp gas and diesel fuel and smog all mix together to produce a city's night aroma.

The Grand Hotel, where I put up, was not so grand. I had been turned away from several better ones and here a room was available. The hotel was old, poorly located, dilapidated but functional, with a hot shower and a decent bed, but no real television reception and the only accessible channels in French. It had a noisy but working air-conditioner that produced a steady hum of white noise, blocking out the traffic sounds from the street. The other hotels were filled with official travelers from all around the region, here to attend a microcredit summit.

My first morning in Abidjan was frustrating. It was difficult to communicate. Many calls did not reach their intended party. I accomplished little more than getting an appointment at the U.S. embassy for the next day.

The so-called Paris of Africa was not all that it was cracked up to be. It was crowded and trash strewn, and the most expensive hotels looked on the outside like tenements under repair. The streets were choked with traffic. My guide, Assogba, was nice but ineffectual. I would have preferred a guide-interpreter fluent in English, one who knew the lay of the land.

By noon it was raining steadily, which I hoped would cut the stifling heat. The clouds had been building all morning. Outside my window a little *Prinia* warbler was singing like a tailorbird atop a low mango tree. Pied Crows, beautifully patterned in black and white, soared like ravens over the city. A graceful swallow much like our Barn Swallow swooped over the freeway. The rain seemed not to bother the birds. Abidjan was distinctive for its bird life as well as for its crowds, narrow roads, wild highways, and tall buildings atop hills.

Night in the city: it was dark here by 7 P.M., and I was warned not to
wander around at night on foot. But the thought of eating in the dreary restaurant at the bottom of this dreadful hotel was too much, especially when
I had been told that there were excellent restaurants around. I screwed up
my courage and got the name of a Vietnamese restaurant in Plateau. I spoke
to a cabbie outside who did not recognize the name or location. Bystand-
ers kindly involved themselves in my interesting affairs, and before I knew
it I was in the taxi driving at high speed across the De Gaulle Bridge in the
opposite direction from Plateau. After ten minutes I was in Koumassi, the
gritty, industrial sector on the road to the airport. It was not a place I wanted
to be in the dark of night. But the cabbie pulled into a small side street and
he pointed out a tiny Vietnamese restaurant, its pretty tables under a blue
canvas canopy, open to the balmy night air.

After I roundly abused my cabbie (in English) for taking me on what
appeared to be a wild goose chase, he kindly offered to wait for me rather
than leave me in the lurch in this isolated spot. I dined on crispy spring
rolls and some beef and bamboo shoots, food that was far better than what I
would have eaten in my hotel dining room. I was beginning to like my cabbie again. We raced homeward . . . only to encounter a police roadblock on
the road to the hotel. The cab was waved over and the policeman asked for
credentials and identity. I sat back and watched as my taxi driver was shaken
down. Then the policeman eyed me and asked for some identification. My
passport was in the hotel. I showed him my Maryland driver's license. He
squinted at it and glowered at me, then spoke rapidly to my driver in French.
I waited passively and said a few things in English. He was waiting for a bribe,
which I mutely failed to produce. After a few minutes' standoff he gruffly
sent us on our way. The taxi driver, chuckling, told me what the gendarme
was after. Such was the night life in Abidjan. On return to the hotel, I found
a cockroach perched on the doorknob of my room.

❧❧

I was here in Ivory Coast doing *international conservation,* a field that has
been and continues to be driven by adventurous Westerners like me—thou-

sands of us. Less polite souls might call us enviro-carpetbaggers, whose work and livelihood depend on environmental decay in developing countries. We are enviro's for hire. We put up with mosquitoes at dinner, cockroaches at bedtime, and, commonly, showers of cold or tepid water in the early-morning rush hour. We put up with earthquakes, insurrections, war, corruption, chaos, and confusion. True, we do this in an attempt to save the forests and wildlife and lifeways of the traditional forest peoples of the world, but mainly we do it because it is our livelihood—it's what we do. Our expertise, by some neo-colonial quirk, is not much needed in our own country; so to put bread on our families' tables, we travel the world in search of environmental problems that can attract significant funding by the aid agencies of the world. Lots of politics are involved. Lots of jealousies and competition. What a witch's cauldron of language, culture, science, and politics!

Is this bad? No, just sad and inefficient. We are expensive, and usually we know little about the countries we visit. Our societies have produced an overabundance of overqualified "world experts" on bugs, birds, rainforests, climate change, sustainable forestry, and the like. We are the best, but there should be a better way to solve the problems of the developing world. It is a matter of cost effectiveness. Our ultimate beneficiaries are typically rural poor who earn less than $500 a year. We might spend this amount in a few days in the capital city, trying to figure out which way is up. Still, we have models and systems and hypotheses, all of which can be put to the test in the field with the rural poor and their biodiverse habitats. If we succeed there, we can translocate that model to the next country, and perhaps win a grant from some multilateral institution for our new idea.

Conservation projects succeed if a park is established, or a management plan is put in place for an established protected area, or if a government delineates a network of conservation areas. It's a bit more complicated than it sounds. But the world of "rural development" is ever so much more complex. I thank my lucky stars I am not in that particular field. In fact, on this trip to Ivory Coast I was as close to being an international development expert as I ever have been. For here I was selling Counterpart's *Forest Gardens* system—a tropical permaculture (site-stable village agriculture) that promotes natural forest regeneration. Although *Forest Gardens'* ultimate objective is forest conservation, the process of getting there looks a lot like rural development through innovative agriculture.

Think about how difficult it is for the international rural development expert! Lifting rural populations out of poverty has to be the hardest task known to humankind. We all know there are more rural poor today than ever before. While we also know that some developing nations have gone from desperately poor to somewhat prosperous (countries like Malaysia, South Korea, and Botswana come to mind), it is doubtful that this felicitous change was a product of international interventions in the form of development aid projects. The best way out of poverty is, I presume, for the nation to pull itself up by its own bootstraps (again, easier said than done).

The superficial difference between a rural development project and a conservation project is a big one: the development project seeks to improve community livelihood; the conservation project seeks to protect a piece of nature. But the two can be intertwined, because virtually all conservation projects also require changing the behavior of one or more communities, often communities that are desperately poor. As a result, many so-called conservation projects resemble rural development projects. Because of the widespread global desire for "poverty alleviation," many international funding agencies want their conservation projects also to improve conditions in the stakeholder communities within the zone of influence of their projects. The bar is set very high, and in essence requires the conservation team to do conservation and rural development at the same time. A demanding task indeed! The team will either focus on the conservation objective and not get the development done, or become confused and see the rural development as the "means" to conservation and end up overinvesting in that to the detriment of the conservation objective. The best way to address this conundrum is to establish a partnership in which a development organization does the community development and a conservation organization does the conservation.

By 2008 the aid agencies of the world (World Bank, United Nations Development Program, U.S. Agency for International Development, to name three) were allotting billions for poverty alleviation, but far less for nature conservation. Many nature conservation organizations are therefore driven to try to look like rural development organizations, in order to get at that big pool of money.

Ironically, although the lion's share of aid investment goes to poverty alleviation, the best scientific evidence indicates that communities are poor because they inhabit a degraded environment. By contrast, rural communities that live in a largely natural environment are, in the main, "rich" in terms of most of the necessities: clean water, abundant fish and game, forests rich in medicinal plants, timber and fiber for building, and fertile soil for productive gardens.

Think of the poorest of the poor—those living in the Sahel, the dry area I had just flown over from Paris. People living in those extremely dry environments consume the habitat's natural resources in no time: (1) the fresh water is in short supply because there is little surface flow and the artesian water is quickly depleted; (2) the fragile native vegetation is quickly stripped by overgrazing, mainly by goats and by the villagers' nightly need for firewood; (3) the soil is blown away because it is not protected by windbreaks or natural vegetation; (4) there is no wealth of game or timber to provide noncash benefits; and, (5) crops are at the mercy of the all-too-frequent droughts that strike low-rainforest areas. Most very dry environments simply cannot support human populations for the long term. The people suffer and the natural environment is decimated.

In the early morning I awakened from a bizarre dream to the growl of heavy thunder and the pounding of rain. It was the height of West Africa's rainy season, when even the mornings can burst awake with thunder and downpour. Out my window, people were fleeing to their jobs. I groggily looked out over the fetid lagoon adjacent to the hotel. A feeling of diffidence and fear rose up in my belly—another day facing an unknown and hostile world that speaks a foreign language! Never mind. A hot shower (it was hot this day) and a quick shave, and a breakfast of tea and croissant, got me going, first to the U.S. embassy.

Waiting for me in the hotel lobby was my interpreter-guide, Assogba Felicien, very dear and gentle and trying to be helpful. He had the facial scarification that is so common in this part of the world. Old traditions die hard in West Africa. It appeared that he was enjoying his work as much as I appreciated having someone help me. He was a nice companion.

The embassy inhabited a paranoid State Department universe that came into being after the Al Qaeda bombing of the Nairobi embassy in 1998. The entire block was cordoned off, which was sad news for the business establishments that needed to make a buck near this virtual war zone. Here, though, one could see the true level of West African security. I sat in the high-security lobby, a young and grim-faced Marine peering through the thick bombproof Plexiglas, guarding access to the inner sanctum. But then I watched as a happy brigade of painters with covered buckets passed around the security checkpoint and headed right into the embassy proper without any display of credentials or check of their materials for possible weaponry.

I met with the embassy's environment officer, who gave off distinct CIA vibes. He gave me a military handshake and spoke in shallow generic euphemisms and totalitarian remedies. The agricultural attaché proved to be more my style, and we talked environment and agriculture. These embassy jobs had to be difficult: brief station appointments, new cultures to assimilate and understand, too many crises to address, and inadequate funding because the U.S. Congress hated foreign affairs. Tough, thankless tasks. These embassy people tried hard, in spite of the many bureaucratic disincentives.

I lunched with an old friend from Conservation International (CI), someone working in the front lines of biodiversity conservation. We had lots to talk about, reminiscing about our days together back at CI headquarters, details of the scene here in Ivory Coast, even raising the possibility of institutional collaboration. The shock troops of international conservation were few and usually friendly; we all knew one another.

It rained three times during the day. Yes, this was the rainy season. We saw a touch of blue sky momentarily, but mostly it was brooding gray cloud. A walk down the street in the afternoon was eye opening. At every third or fourth street corner I came upon a garbage heap. For some reason trash was dumped at these particular sites—no indication of when it would be picked up. I tried hard not to breathe the heady stench when navigating gingerly past these foul mounds. Still, the shops of Abidjan were varied and vibrant. They had Western consumer goods in abundance, news stalls with an array of continental magazines, and shop after shop of kaleidoscopic West African textiles that were rainbows of design. All the same, the pallor of decay was everywhere.

I took dinner at a lovely maquis—a local West African restaurant with a species of cooking much like the north Indian oven-baked tandoori. I dined with Lorraine Lathen, an engaging African-American woman from Wisconsin. She lived here with her ten-year-old son, working on health education. She assured me she was enjoying her sojourn in this country. The maquis served dry-baked chicken and fish dishes, each with a spicy garnish of onion and tomato, as well as attieke, a ricelike side dish made from pounded manioc (cassava). This maquis was inexpensive and welcoming, and I enjoyed sampling the local fare.

I woke up to dark overcast and heavy rain. No thunder. I breakfasted on tea with milk, breads (croissant, brioche, baguette), and fresh papaya with lime. Things had started to go easier now that I had come to fret less about the language. The time had come to visit the countryside.

Doumbia Sory, my backcountry guide, and Zakariah, the driver, met me at the hotel and got me organized in the tour van. We took off at breakneck speed on the major highway to Yamoussoukro, the capital city. We raced down a dual-lane highway heading north. As evidenced by this grand highway to nowhere, the megalomania of Ivory Coast's first president, Felix Houphouët-Boigny, seemed to know no bounds. This father of the nation took it upon himself to invest great swags of money on every conceivable infrastructure project, especially in and around his tribal homeland a few hours north of Abidjan. It was Houphouët-Boigny who decided that Yamoussoukro should become Ivory Coast's capital, and he set about making it a city with all the candy: Africa's largest cathedral (modeled after Saint Peter's

Young firewood sellers along a roadside in Ivory Coast

in Rome), an elegant palace and grounds, wide boulevards and tree-lined streets, a moat with crocodiles to protect his palace, a parliament building, a massive structure entitled the Foundation for Peace, and a four-star hotel with a top-floor restaurant serving elegant meals in the French style. Even a great highway linking the first city of the country, Abidjan, to this spanking-new capital city.

The first two hours took us through the former rainforest. Lots of large trees were still around and some bush, but mostly regenerating scrub—rich scrub at that. The tall remnant trees were marvelous rainforest giants. Closed forest could return here in fifty years if the people would go away and leave it alone.

As we moved north, it gradually became drier with more savanna-like vegetation. We stopped at a small village where a wonderful, unruly colony of Village Weavers was active, the male birds building nests, or singing from their completed nests, or flying about excitedly. Weavers are relatives of the endearing but lowly House Sparrow, common in most of the larger cities of the world and widespread around farms in the United States. Whereas the House Sparrow builds an untidy nest of loosely gathered materials, most weavers follow their name and actually weave a globular nest from carefully collected strips of grass. In each weaver species, the male constructs a nest that is identifiable by size and shape. Some species build a huge communal structure like an avian apartment building. All show remarkable weaving skills that defy logic. How can a bird be so accomplished with just a bill and two feet?

Equally remarkable is that weavers aggregate to nest in tight and noisy colonies in low trees. These colonies become a virtual mayhem of sound, color, and motion as the males come and go in their nest-building, and then become masters of self-advertisement and display as each competes to attract a mate to its nest. The male Village Weaver is a rich yellow with a black-and-brown face and cheek, whereas the female is dull streaked brown and gray. Village weavers nested in many of the little villages I visited in Ivory Coast, usually choosing a solitary tree right in the middle of the village. The noise the weaver colony produces is amazing, and the perpetual activity is in some ways analogous to an African village. The birds are ignored by the village residents, except for the occasional mischievous boy who tries to bring down a bird or two with a slingshot.

The three of us took lunch at a huge maquis in downtown Yamoussoukro, within sight of the basilica. Loud music was playing, and we sat high under a roofed structure with no walls, able to look out onto the artificial town lake. The scene was picturesque but the lake was, thank heavens, far enough away that we could not smell its fetid water. We ate broiled chicken and a whole baked fish with the obligatory garnish of tomatoes and onions. We had a whole range of side dishes: fried potatoes, fried plantains (cooking bananas), attieke, rice, and the wonderful side sauce that was red with tomato and in which lurked secret spices, reminiscent of Jamaica jerk seasoning.

Another hour's drive on a smaller road took us to the town of Bouafle, gateway to Maroué National Park, a mix of gallery forest and woody savanna that was home to elephants, hippos, and crocodiles. I was in Bouafle to visit the Conservation International field office. Carla Short and P. J. Stephenson were running CI's European Union–funded program to help manage this particular park. In an afternoon meeting with the CI team, I made the case for a collaboration on the project, in which Counterpart could focus on village agricultural stuff and CI could continue to focus on forest biodiversity. Everyone liked the idea. We proposed to discuss it at length the following day.

Late that afternoon I visited Carla in her nice bungalow for a beer (she lived in the neighborhood the local people had named Millionaire's Row). At dusk my little party of three drove 20 miles up the road to the Park Hotel, the last leg being a dirt road in the pitch black of night. Given that we were in the countryside next to a national park, I was surprised and disappointed that we saw not a single living creature along that 8-mile stretch—not even a snake or nightjar. Apparently the local people were efficient hunters.

Safe in my little grass-roofed hotel chalet (named Elephant) I thought back on all the forest remnants and the roadside villages I had seen. The challenges of sustainable development here were huge. The villages were poor and overcrowded. Tiny children abounded. There was so much to do. Could we tackle something so large?

It was a night of lightning, thunder, and rain—a tropical wet-season evening. I slept heavily until the alarm went off at 5:45 A.M., well before first light. Doumbia knocked on my door shortly thereafter and we drove down the entrance road into the park, looking for wildlife. A very light rain fell, although much of the sky was blue. The park was quiet, more quiet than we had hoped. We encountered an African Pied Hornbill, then pairs of them, moving through the canopy, calling their high-pitched piping notes. Then we glimpsed a Red-billed Hornbill, small and paler, again in the canopy. We flushed some birds out of the roadside grass and thicket, but these blundered off unidentified. A small mongoose sat in the road, looking at us inquisitively. A Green Turaco appeared, lustrous green, with wing feathers that flashed blood-red in flight. The turacos, with their plumage painted in the palette of ripe rainforest fruit, are Africa's most beautiful bird family.

The park road passed through a mix of savanna and open forest, dominated by massive kapok trees, rumored to be the home of forest genies. These lovely trees were truly distinctive, with their striped bark and sinuous trunks bulging at the base. Tall green grass grew thick by the roadside—in which hid green mambas?

We saw little evidence of mammals—few scats on the road, and just a single mongoose observed. Birds were not terribly abundant, but novelties appeared in sufficient quantities to keep us alert. Still, it was apparent that this park environment had been much impacted by humans through fire and deforestation, and that the biota was both wary and thinned—the result of chronic hunting by the local residents.

We returned midmorning to our quaint hotel for breakfast. Upon leaving, we stopped for some roadside purchases of plantains and yams (for the men to take back to the city, where these things were much more expensive), then to the CI office for more discussions with P. J. and Carla. After satisfying talks and promises of collaboration, my little team headed into the countryside to interview villagers about their livelihoods and agriculture. Local people's lives here revolved around agriculture, a combination of subsistence and cash crops. Getting that balance correct was difficult, and managing the crops around the dry season and its relentless fires was a challenge.

Another huge issue was the transnational movement of displaced people. People seemed to move at will among the former French colonies of West Africa. "Unoccupied" lands (often parks) were targeted by settlers from other areas (even other nations), which made nature conservation and sustainable land management difficult to achieve. It was an issue entirely foreign to a place like the United States or Papua New Guinea, where land tenure remained strong. Here, tribes rich in land were generous and allowed less fortunate landless groups to occupy areas that were undeveloped. These typically were the areas most important for biodiversity.

In the late afternoon we drove, through heavy rain, from Bouafle back to the remarkable and grotesque Yamoussoukro—with its great basilica and its towering Hotel President, where I was to stay. Yamoussoukro was an African version of Canberra or Brasilia, with huge edifices and wide streets with zillions of streetlights, but not many people, many undeveloped lots, and the impression of incipient decay. Still, it was a much more pleasant town than Abidjan, with some excellent maquis.

The next day we departed early for the Reserve d'Abokouamekro, about 40 miles to the east of this vast, boulevard-gridded town. About 5 miles from town the road degenerated to puddled red mud. Our speed dropped to a crawl to dodge the largest of the potholes. High grass walled either side of the road, and tall kapok and other jungle trees loomed from the grass in the near distance. We passed through big villages, squalid and poor but with brightly painted concrete houses. Goats and children were everywhere, also bare earth and rundown markets. I could imagine how unpleasant these communities would be at the height of the dry season in December. Now, in the rains, they were probably almost livable.

We passed through a gallery forest and made a stream crossing, where the previous afternoon's downpour had evidently produced a flood. All the tall grass had been thrown down by the heavy floodwaters in a swath about a half mile wide. We three agreed we would return to the relative safety of Yamoussoukro before the afternoon rains got serious.

The reserve encompassed two large savanna-topped hills. Forest nestled in the deeper ravines. The fertile farming land around the perimeter of the fenced reserve was lush and productive.

We saw some men in a yam garden and approached them to chat. They were five hardworking, barefoot souls. It turned out they were a cooperative work group. The five worked as a team, planting each man's individual garden on successive days now that it had rained. One man served as our informant, and he answered our many questions with clarity and eloquence. We learned that his chief crops were yam, corn, manioc, plantain, and cacao. The first four served for subsistence, with the excess production sold at market. Cacao was the cash crop. For yams, he planted one crop a year. He stored his own rootstock (which we saw in yam sheds). For meat they all depended on products of the bush—agouti (bush rat), monkey, whatever they could trap. They told us they loved bushmeat. An El Niño fire the preceding year had destroyed much of their cacao, so they were replanting. They liked the adjacent reserve in theory, but its presence did pose problems. It had closed off access to an important vine they used for yam storage. Also, their ancestral burial grounds were in the reserve; they were no longer allowed to visit, and it pained them. These farmers had never seen the elephants that lived in the reserve. While the elephants had never caused them direct problems, the monkeys that inhabited the gallery forests in the vicinity (outside the reserve) were a real nuisance and raided their crops. Whenever possible, the men killed these monkeys and ate them with gusto.

We visited a garden house and spoke with some women there, who greeted us with giggles and smiles. They were cooking a soup of greens. At least in this season, life here seemed pretty good. These people were not immigrants; they had lived here for several generations at least. Our main woman informant noted that the area had changed little environmentally since when her father was a boy—it was not being noticeably devegetated.

We saw some wildlife but not a great deal: many Common Pied Hornbills, both turacos (Plaintain-eater and Green) and several hawks, a squirrel, and a mongoose. We encountered a lovely green snake that might have been a green mamba—fast and slim and beautiful, crossing the road.

The reserve seemed well established and the people living around it were managing; there did not seem to be need to develop a buffer management program. The fence ensured no encroachment. I saw greater opportunity in the national park working with CI rather than here with the reserve.

I learned in my interviews of the CI staff and the residents on the ground that there was a great deal of movement of peoples in the region. Tribal groups were migrating across West Africa in search of economic opportunity, in part because of the environmental collapse of the places they had formerly occupied. This sort of random movement was unfortunate for the local environment and certainly was dangerous for the political stability of the nation, in that it can blur any sense of nationhood.

From a conservation viewpoint, the continual movement of landless settlers threatens every natural habitat. What is one day a pristine park is next day the homeland of squatters clearing forest for cacao and vegetables. Strong and stable land tenure, such as we find in New Guinea, is much preferred by conservationists.

We dined early at the Maquis de Jardin in downtown Yamoussoukro. A Guinean minstrel performed with an unusual multistringed instrument, and

A productive vegetable garden in rural Ivory Coast—productive but not quite
my idea of a *Forest Garden*

the food was awesome—especially fine basmati rice and wonderful chicken
with vegetables.

Later in the evening, we drove out to the barrage of the Koussou River,
an enormous hydroelectric dam. On the drive out we saw lush corn gardens and passed through the cacao and *robusta* coffee plantations of the
president's family (these were full-sun plantations, entirely lacking the natural
shade that is so beneficial to the environment). A young man beside the road
offered a large rat for sale. It turned out not to be the famed agouti (bush rat)
but instead was a serra (which is smaller).

In passing, Doumbia (my guide, savant, and comedian) assured me
that he had to appease the animist gods regularly to ensure success in his
life and career. He had a chicken sacrificed twice a month. Black magic was
here to stay, he assured me. More than half the country followed Islam, but
animism permeated life.

The next morning at the hotel, I gazed out the window of my room at the fluttering forms of the little white-rumped swifts racing by in small parties —so vital and animated. Below was an ornate three-part pool complex with lots of children and parents taking the waters, and a wonderful garden and forest plantings in the background. From my high window it had the look of some earthly paradise. My room in the Grand Hotel of Abidjan was starting to seem like squatter's quarters.

What I had learned from my trek into the interior was that agriculture was king, a fact that was reassuring to Counterpart and its *Forest Gardens* system, with its promise of sustainable agriculture. Permanent crops were appreciated. And the land was fertile. Cacao could be an anchor product around which a green agriculture project could be constructed.

The next stage of our auto adventure took us to the coast. We had planned to travel from Yamoussoukro to Taabo Dam, then to Parc National d'Assagny. We took off in the morning sun and drove down the highway, stopping at a rural market to photograph people and their garden produce. Especially the women here did not like being photographed. Those who did consent had to be financially compensated. The market was bursting with mango, avocado, and yam, with some chilis and a few other things—abundance, but not much variety.

We continued on to the dam of the Bandama River at Taabo. It was more impressive than Kossou, with a huge channel cut straight through granite bedrock to concentrate the water for the turbines. The channel was about a mile long and some 150 feet deep. Another incredible human feat that demonstrated Houphouët's ego at work again. Swarms of Little Swifts and Lesser Striped Swallows were playing around the spillways. The reservoir lake was picturesque in the rainy season.

After sightseeing at the dam, we zipped down to Abidjan, veered west at town's edge, and made our way toward the Parc National d'Assagny. On the way, we passed through huge monoculture plantings of oil palm, rubber, banana, and even manioc. Rain fell in buckets for more than an hour, a prodigious tropical downpour. Following the signs to the park, we turned off onto a tiny piste (track) into equatorial jungle. It was dark and forbidding, made all the more unwelcoming by the overcast sky and intermittent rain. Midjungle, we turned a corner to find that a large tree had fallen across the road. The men began cutting the tree with a bush knife while I hiked the nearly 2 miles to the park hotel to alert the staff. I walked briskly for a half hour through dank forest with the occasional small clearing. No wildlife was to be seen, only silent and motionless forest, lovely though it was. I spotted only a tiny family party of tailless warblers.

I walked up to find the hotel abandoned with not a soul around and the grass growing high on the front lawn. The entire facility was in poor shape and the front doorway was ajar, open to the elements. It looked as if the property had been abandoned in a big hurry. Why was the front gate at the park boundary not closed? It all gave me a strange and creepy feeling.

On my hike back, I was met on the road by my car. We drove down the entrance road to the park hotel I had just visited. Still nobody there, and it was supposed to be our night's lodging. Throwing up our hands, we turned around and began to develop a fallback plan. We proceeded to Grand Lahou, through the wasteland of oil palm monoculture. Every scrap of rainforest was gone, except that in the park.

Grand Lahou gives onto a lovely sawgrass-and-palmetto wetlands plain much like the Everglades in Florida, but enlivened by flights of several species of hornbills in small flocks. It was late afternoon and these groups were headed to roost. I saw at least a hundred hornbills on this evening, even at our motel back at the edge of Grand Lahou town. I had been seeing hornbills all afternoon since leaving the park. Plaintain-eaters were common also. The best sighting in these everglades, however, was the strikingly eagle-like white-and-black Palm Nut Vulture, attracted by the abundance of fan palms that dot the wet prairie. This weird bird flies with a slow grace much like an American Bald Eagle, but sports a very short tail. Upon first inspection I thought it was some large goose, then an eagle, and only after checking the bird book did I realize it was definitely this vegetarian vulture.

We had settled into the only motel in the area, just south of town. It was tiny and quiet and empty, with six rooms and a bar and a patio restaurant. We took a tasty meal of fresh seafish steak with rice, attieke, pommes frites, and the obligatory tomato-onion sauce, started off with an avocado salad with vinaigrette. And of course a small Flag beer. Our peace was undisturbed by any overhead airliner or other city sound, just the churrs of the katydids. Mosquitoes were out in abundance as well. Because of the rain and clouds it was quite cool; I wore slacks and a sweatshirt.

This region of coastal Ivory Coast had been devastated by large-scale plantation development of rubber, oil palm, coffee, corn, and manioc. The roadside villages showed no evidence that the large-scale agriculture was providing them any benefit at all. At least the everglades wetlands had not been converted.

After dinner I spoke with a teacher who had spent some months at a summer camp in Michigan years before. He liked to work on his English so we chatted for an hour. He told me that in preparation for retirement he had planted 27 acres of cacao and coffee. The coffee was producing well but the cacao was doing poorly. He was supposed to apply fertilizer but he had no money for it. Nor did he apply pesticide. He wanted help with the ailing cacao.

It had been quite a day, from the sunshine of Yamoussoukro to the beating rain of Dabou, to the pleasant dusk of Grand Lahou, and then the nighttime serenade of insects on the outskirts of the coastal everglades.

Monday morning we awoke early to return to the everglades. Again the weather was miserable—heavy low clouds and drizzle. Still, we persevered. The clouds scudded low over the wet grasslands, while scores of hornbills passed by. My birding produced another Palm Nut Vulture, plus an assortment of novel species (like Orange-masked Waxbill, African Pygmy-Goose). The best thing about the morning was simply our presence in this lovely environment. It was bucolic and undisturbed, dotted with lovely palms everywhere, views of the Bandama River across to Parc National d'Assagny's forests, all quite wonderful. The weather gradually brightened over our two-hour trek.

Afterward, we rushed back to Abidjan for our scheduled afternoon meetings. I stopped the car several times to photograph the agricultural degradation along the road: principally rubber, palm oil, and banana. The scale of the monoculture was awe inspiring. The condition of the villages nestled right in the middle of these big plantations made me shudder—garbage, poorly clothed children, bare hardscrabble earth, awful huts.

Back in Abidjan I met up with Assogba Felicien, my city guide. We worked the phone, then did the meeting circuit. I first met with Dr. N'Guessen M., director of environment, who offered official support for a Counterpart program in the country. He promised he would write us a formal letter of invitation. (It also appeared that he was totally overworked.) Then I met up with Monsieur Amani (director of the Department of Parks and Nature), who took me to see the minister of environment, Jean Claude Kouassi. After waiting two hours for the minister, we had a productive twenty-minute meeting, in English. He lectured us on the possibilities offered by husbandry of agouti, bore wells, irrigation, agricultural waste reprocessing for fuel, fast-growing fuelwood trees for villages. There was a lot that interested the minister.

Returning to the Grand Hotel, I now had a room on the fifth floor, which I discovered was where the better rooms were. I had a room service dinner and collapsed in bed early.

My last day in Ivory Coast was a busy one. I was up early in preparation for the morning's field trip with M. Balle of Côte d'Ivoire Ecology (a local NGO, nicknamed CI/E). It started poorly, as there were mixups about car and driver and meeting place. Why this happened was a mystery, as we had called and left messages twice the day before at Balle's office. We did manage to get together in spite of the confusion.

M. Balle and his colleague Mme Danielle and driver took me and my guide in their Toyota Hilux four-wheel-drive to their field site, about forty minutes' drive east of town, at the verge of a vast palm oil estate that formerly belonged to the government (and which was their research station for oil palm). Here beside a large lagoon, a small village eked out a living in the shadow of the vast oil palm estate. The villagers had little land, and had deforested virtually all of it. CI/E had, over five years, installed 10 acres of acacia agroforestry that was now an acacia plantation with much natural regeneration below it. Birds and monkeys used the acacia tract, too. I walked the plot and was impressed. It needed to be moved to *Forest Gardens* status, otherwise the head man would convert it to oil palm, so he told us. The villagers would cut down the acacia plot, even as they noted how wonderful the plot was because it had encouraged the return of a number of medicinal plants they harvested monthly for local use! Such were life's contradictions. M. Balle was skeptical that the villagers would accept anything less than maximum production on their land (hence he was doubtful about shade coffee and shade cacao) and he believed that Ivory Coast had lost all of its shade-tolerant cultivars.

There was real opportunity here, but it would have to be pursued vigorously, and soon. Balle had plantings like this in twenty-five communities around the country, so there was a platform on which to build. On the other hand, he freely admitted the limited capacity of CI/E mainly because of financing problems. He had cadged a United Nations car on loan for a project ("UNHCR" was written on the door). He had been director of the national agronomical research institute, which had now been abandoned altogether! No more forestry research was going on in the country, either, which M. Balle heavily lamented.

As a naturalist, I was pleased to walk into Balle's little piece of home-grown acacia jungle. There I found a paradise flycatcher and a pied barbet. It is amazing what five years of work can produce on once-cleared land. A community forest scheme might well succeed on its own, with each community regrowing its own forest plot of 10–20 acres.

In the afternoon I meet with Peace Corps director Sachiko Goode, a cheerful American woman who had been directing the office since 1995. She was open to potential collaboration, through either the health program or the environmental program. We had a nice chat, after which she loaned me a car and driver to go to the World Bank.

There I visited Jean Michel Pavy, chief of environmental programs, who was apparently *the* man when it came to environmental aid programs in this country. He first told me that he hoped he had rescued the bank mission of a month earlier that was scoping out the $40-million protected-areas program for Ivory Coast (similar to the Madagascar protected-areas program). Apparently my new buddy, Minister Kouassi, had not signed off on it; he claimed ignorance, throwing the bank team into a tailspin. Pavy was a delight and expressed interest in *Forest Gardens.* (It is always the hope of a fledgling NGO program to attract the attention of the giant World Bank.)

Assogba Felicien proved his mettle that evening at the Felix Houphouët-Boigny International Airport, just twenty minutes from my hotel. Assogba appeared instantly upon arrival of my car and whisked my bags on a trolley toward the check-in counters. He had me processed at high speed through a series of gates and checkpoints, and I was deposited at the waiting lounge behind the immigration barrier. His ability to move in and out with impunity was the payoff for my making a local friend through trust. Only at this point did Assogba inform me that he was from Benin, not Ivory Coast! He had been living in this country for seven years, having arrived to join his uncle and work. But things had turned out poorly. Thus his rather menial and informal job at the airport, where I had encountered him a fortnight before.

This chapter has focused on the process of rainforest conservation. *Process* is so much a part of international nature conservation; it involves meetings, travel, fact finding, and planning for events that many times never work out. This realm of conservation in the developing world has considerable slop in the system. It is not for the faint of heart or for the perfectionist.

Ivory Coast—what a lovely, messed-up place! What an Africa mélange of good and bad! On the one hand, I saw heaps of garbage piled in the streets of Plateau, Abidjan's nicest section. On the other, I spent time with wonderfully kind and helpful people like Doumbia and Balle and Assogba. I encountered superior bird life, such as the turacos and the abundant hornbills. But I also noted the missing mammal life that had fallen prey to the bushmeat trade. It was a land of contrast: desert in the north and wet rainforest in the south, the lovely everglades of Grand Lahou surrounded by oil palm and rubber monocultures.

This exploratory trip to Ivory Coast seemed to encompass an eternity. Each time I depart my home environment and am enveloped in an entirely new and strange overseas environment, time seems to break down. I become detached from all the familiar objects of home. The trip feels much longer than it really is, and the return becomes that much more disorienting.

Part of the illusion is brought on by the intensity and unrelenting nature of the work: day and night meetings and travel and need to focus on the new and unfamiliar environment. It is exhilarating but mentally exhausting. Arriving home in Bethesda to find the routines of normal life in place makes me see how different much of the rest of the world is from the United States.

This particular trip reminded me how important it is to visit new environments, where all of our faculties are tested and our consciousness is stretched to the limit. The intellectual challenge teaches us so much about the diversity of the world and reminds us of how parochial our home life is. It is crucial for each of us, from time to time, to break out of the home cocoon and metamorphose into a more worldly-wise, open-minded person.

If America wants to re-earn its cloak of global leadership, its leaders must develop a generous and open worldview that only can be gained through travel to the little-known places, where other languages are spoken and other foods are eaten and other gods are worshipped.

❧❧

So much has happened in Ivory Coast since my visit there in 1999. The nation as I saw it no longer exists because of constitutional crisis and revolution and bloodshed and chaos and de facto partition of the country by warring factions. The Conservation International team has long since departed, its project and its office shut down. One can only wonder what has happened to the Côte d'Ivoire Ecology and M. Balle? What of the minister of environment? What of my guides and driver? Are they still alive? Are they safe?

Political uncertainty is one of the enormous burdens of the field of nature conservation in the developing world. Compare the objective of a logger to that of a conservationist. To succeed in his business, a logger needs to gain access to the forest, harvest the timber, and depart. The logger succeeds if he sells the harvested timber for a decent profit. By comparison, the conservationist must protect that old-growth forest for *all time.* A single wildfire, one large-scale bout of illegal logging, an invasion of landless migrants, a change in government that leads to revocation of the park mandate, all spell failure.

Has a conservationist succeeded if he has protected a forest and its wildlife for fifty years, and it is logged out in the fifty-first year? No. That's why conservationists have trouble sleeping at night.

At a more general level, it is obvious in hindsight that the societal and political currents flowing in Ivory Coast in 1999 were not conducive to long-term nature conservation. There are certain political pillars that support conservation: good governance, stable human populations with clear tenure rights over resources, freedom from corruption, the absence of extreme poverty. Apparently none of these conditions are present in Ivory Coast today, and that is why conservation there is a long shot at this time.

National strife and political disintegration are two major forces that threaten environmental projects throughout the developing world. Those of us in the field work hard and plan in detail, and our institutions invest hard-won funds, only to see at times the obliteration of everything that has been built. It is the high-risk nature of our profession.

Although I am now based at Conservation International and focus on the Pacific region, the riskiness of the work continues. In 2006 friends and colleagues died in a helicopter crash in Nepal; twenty-four passengers perished. I have seen a project in Papua New Guinea threatened by the political machinations of an angry provincial governor. We work with risk every day and come to accept it. Yet when something unfortunate happens, we find it shocking. On a day-to-day basis, our brains force us to block out the ever-present threats. It is what we need to do in order to work effectively in the face of such challenges.

CHAPTER SEVEN

Local People Really Do Count

In the jungle the going was fairly easy.

High overhead, two or three hundred feet,

the treetops rested over everything like a blanket,

shutting out the sun. It was cool and dark and

not so warm that walking was uncomfortable.

It was very silent, this forest, but not gloomy.

Dillon Ripley,

Trail of the Money Bird (1942)

IN 1982 I HONEYMOONED WITH MY wife, Carol, along the Sii River in the Lakekamu Basin of Papua New Guinea. I was there to survey birds of paradise. My bride was there to see, first hand, what her new husband did for a living. The basin is a large expanse of lowland forest in the center of this western Pacific nation, physically protected from imminent exploitation by a ring of surrounding hills and mountains as well as some rather imposing rivers and swamplands.

Carol's grandmother, not surprisingly, was doubtful that our brand-new marriage would survive a honeymoon trip to the jungles of New Guinea. As a result, she promised us a two-part wedding gift: $500 upon our marriage, and a second payment of $500 if we returned from the jungle still married. Carol and I collected both installments, and the rest, as they say, is history.

Eleven years after my honeymoon visit to the Lakekamu, I returned to the basin with a group of research students to initiate field studies and to develop a conservation program there. This I was doing under the auspices of my two employers at the time, the Wildlife Conservation Society and Conservation International.

June 15, 1993. I arrived at Jackson's Field, Port Moresby, in the early afternoon on the Air Niugini flight from Brisbane, Australia. As usual, I had made the trip from the United States without a break. I am a believer in not laying over on this long trip (Washington–Los Angeles–Brisbane–Port Moresby), which usually entails about thirty hours of transit time (air time plus airport time). I have made this trip forty times in the last thirty years, and have developed a tried-and-true methodology for surviving with minimal damage: sleep at least ten hours on the trans-Pacific leg of the trip and, after reaching the destination, don't sleep until after dark, no matter how tired or disoriented.

I settled into the dowdy Dove Hostel in Boroko, a Port Moresby suburb. Run by one of the many church denominations in PNG, the Dove was a five-minute walk from the best concentration of retail shops in Port Moresby, and thus was a convenient spot from which to stage an extended field trip into the rainforest. In the early 1990s the Dove was my favorite place to stay in Port Moresby because of its strategic location and its cost—a mere twenty dollars a night. That afternoon I managed to fight off my jetlag by spending

Preceding page: Villagers swarming our Cessna at Kakoro Village

time in the shopping district, but could not purchase anything because I was unable to exchange dollars for kina (the local currency). I also visited the University of Papua New Guinea (UPNG) and the Foundation of the Peoples of the South Pacific (FSP), my two main institutional partners for our Lakekamu initiative.

I had planned nine days in Port Moresby, much of which was to be devoted to conservation planning meetings with the government, partner nongovernmental organizations, and colleagues at the university. Because I was making the transition from pure research to research-and-conservation, I now had to spend more time in this unprepossessing capital city. Conservation is largely communication, negotiation, and politics. That had to take place mainly in Papua New Guinea's towns and cities, as well as in jungle communities.

My Port Moresby stay was punctuated by morning meetings, lunchtime meetings, and dinnertime meetings. After my second day in the capital I learned that I had to move out of the Dove Hostel; it was filling up with missionaries, who had priority access to the rooms. I shifted over to the guest house of the Australian National University, called the ANGAU Lodge, another low-rent place to sleep. Field biologists are incredibly tightfisted and habitually stay in the cheapest lodgings. They can stand just about any conditions; after all, they are used to sleeping on the ground in the forest.

Port Moresby at the moment was cool, a pleasant change from the norm here in June, the end of the rainy season. The evenings were breezy with a sweet sunset, then a cobalt night sky filled with bright stars. Still, the afternoons tended to be dusty and sweat-producing, especially when I was fighting traffic in an ancient Suzuki mini-pickup, rented from my friends at the FSP. The alternative, however, was brutal: lots of walking, standing at bus stops, and bumping about in crowded public transport on the potholed roads of this awful town.

In 1993 Port Moresby was on the verge of developing a terrible reputation. For decades it had been dusty and insubstantial. Starting in the 1990s it became dangerous, with random violence being committed by "rascals" against law-abiding citizens. Today it is thought to be one of the most dangerous tropical cities in the world—a major hardship post for foreign service officers and international development bureaucrats.

My fourth evening in town I spent at David and Yasmin Vosseler's flat for a dinner of lasagna and the chance to watch a basketball play-off game from the States (they had a satellite dish). I had to depart midgame to pick up Francine Wiest at the airport. A recent Harvard graduate, Wiest was one of this summer's field students.

The ANGAU Lodge housed quite a crew; besides Francine and me, it included Marc and Bonnie Thompson (from FSP), David Boyd (from the University of California at Davis), and a British anthropologist who quickly convinced everyone that he was an absolute madman. Lots of scurrilous field stories were recounted over beers savored on the tattered couches in the living room. After a day orienting Francine to Papua New Guinea, we visited the Sizzling Steak House for dinner and found we were the only patrons. The place, in the back of a Boroko shopping center, was run by Chinese and served Chinese food, but made the rather curious claim of being a steakhouse. Our dinner was reasonably good, so we were left wondering why the restaurant was not crowded. Rumor had it that this place was a front for some sort of illicit activity—drugs, smuggling?

On Sunday morning I rose well before dawn to take Bob Kerr, the Peace Corps assistant director, up to Varirata National Park to see the dawn display of the Raggiana Bird of Paradise. Our rather unlikely link was that, a couple of years before, Kerr's wife and my wife had delivered babies two days apart at Shady Grove Adventist Hospital, in Gaithersburg, Maryland. Also, I had been in discussions with Kerr about getting Peace Corps volunteers out to the Lakekamu Basin to support our conservation project there.

After the winding hourlong drive up onto the Sogeri Plateau and the park entrance, we hiked in to the creek-bottom display tree to observe three orange-plumed males strut their stuff for a dozen or more females and sub-adult males. I had spent many days observing the happenings at this lek tree in the late 1980s with Jack Dumbacher and others, but we had gotten sidetracked when Jack discovered the chemical secret of the poisonous Hooded Pitohui. Today I would probably not take a chance visiting Varirata

National Park because of the crime along the road. Dozens of bad incidents have occurred over the past decade or so, obviously diminishing the place's friendliness. On this day in 1993 we were all quite satisfied to spend time watching some of the world's most beautiful birds shaking their booty for the ladies. The raucous cawing, the shimmering orange flank plumes, the unearthly postures of these birds of paradise, and the rich greenery of the ravine forest made for a memorable morning.

Monday broke cloudy and windy, with a sprinkle of rain. It was our last day of preparation for the summer's field trip. Francine and I raced around town in search of field supplies. As it was Father's Day (because it was still Sunday in D.C.), I called home and spoke to my wife. Carol and our baby were at my in-laws' house for the celebration *sans* daddy. Being away for many holidays is a hardship for tropical field biologists and field conservationists, only partly assuaged by long-distance calls.

Tuesday we were up at 4 A.M. to ferry our supplies to the airfield. Then I picked up two of the field team: Robert Bino (from the university) and Christopher Unkau (from the Department of Environment and Conservation). After a weather delay, we filled the lumbering twin-engine Norman-Britten Islander and flew to Kakoro airstrip, about fifty-five minutes northwest of Port Moresby. The flight was cloudy, ruining our chance for aerial photography of the spectacularly rugged scenery below. Most of the flight from the capital to Lakekamu is over unbroken forest.

The Kakoro airstrip and its adjacent village were cut from tall jungle on a flat patch of floodplain that overlooks the tumbling white water of the Biaru River. The plain is some 250 feet above sea level, just at the foot of the outermost reaches of the central mountain range, which in this area tops out at 9,000 feet a mere 10 miles north of here. Kakoro, which serves as an administrative center for the local district, is on the eastern verge of an ancient interior lakebed (long since dry) that is now about 150 square miles of flat lowland rainforest, swamp forest, and herbaceous swamp lands. This basin is a naturally protected forest reserve, with about fifteen hundred inhabitants scratching out a living around its edges (the hilly terrain is more livable because it is less swampy than below, and the soils are better for gardening). But

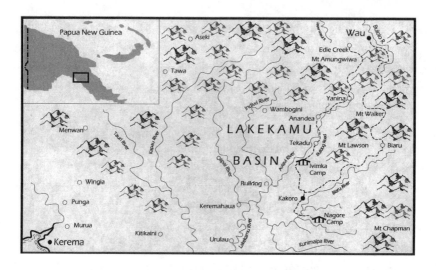

the flat basin forests are rich in game (cassowaries, wallabies, the powder-blue crowned pigeons) as well as alluvial gold. The Lakekamu Basin was home to one of Papua New Guinea's first gold fields, and local prospectors continue to work the streams that "show colors."

Our arrival at the Kakoro airstrip was greeted by a surging crowd of curious villagers. The droning of the twin-engine craft had alerted the community to our approach; given the lack of competing entertainment, few could resist coming to see who was arriving at this lonely destination. Airplanes here are rare as hens' teeth. Kakoro was located in this isolated spot because it was a reasonable place to put a landing ground. Nobody had lived here before the airstrip was constructed, and the peculiar geography of its placement created not wholly harmonious interactions among the four ethnic groups that claimed parts of the basin. Overall the basin had been something of a no-man's-land, with only tiny scattered communities and no government services. This was, in part, because the boundaries of three provinces met here (Morobe, Central, and Gulf), and the first two of these apparently would rather pretend that this land was not theirs. Thus, by default, the entire basin was treated as belonging to Gulf Province, a fact that I learned the hard way in the 1990s.

I had selected this site for a research-and-conservation program because on the map a large swath of beautiful lowland jungle was clearly situated in Central Province. I had wanted to work with the government of that province on conservation, because its offices were in Port Moresby at the time, which made communication very convenient. By contrast, the capitals for Morobe and Gulf provinces were distant from the basin and from Port Moresby.

That the entire basin was managed solely by Gulf Province was the worst of all possible worlds for several reasons. First, Kerema, the capital, was a little swamp-infested town as isolated as any in Papua New Guinea. Second, the politics of Kerema and Gulf Province were not unlike the swampy politics I had read about from the 1920s Louisiana of Huey Long. Third, the ethnic group I found most collegial happened to be the group in least favor in the halls of power in Kerema. The Kurija-Kunimaipa were mainly a Central Province tribe and had no influence in Gulf Province.

On the Kakoro airfield we were greeted like returning heroes. It was my fourth expedition to this fascinating forested basin, and I was known to many. People called out my name and sought to shake my hand. The crowd happily picked up our several hundred pounds of supplies and lugged them the half mile to the health center. Our small team was planning to sleep in the health center main hall (now empty of beds) for the night, to await our dawn departure for our field camp 6 miles distant on the banks of the Nagore River.

Some of my field students had been working at Nagore Camp since April, so our little contingent was to join up with a team already at work in the forest. Although the floor of the health center was quite hard, the evening's rain provided a steady thrum of white sound that lulled me to sleep, with dreams of fieldwork and new discoveries.

On June 24 we finally headed into the bush. We were all up early for a breakfast of navy biscuits and sweet tea, and we sent out the request for porters to carry our huge mound of gear and food the 6 miles to Nagore Camp.

The morning broke foggy and with considerable gloom, not unusual in this piedmont location, where the humid winds continually push up against the mountain wall to the east. I saw scattered individuals standing about in the fog, shifting from foot to foot, waiting for the designated time to grab our cargo and carry it to Nagore. I needed fifteen porters, each of whom would carry about 30 pounds, the standard bush load.

We were off by 7:30 A.M. The trail first led up by Kakoro hamlet, thence down along the banks of the Biaru to Mirimase Village and to the Biaru River crossing below Mirimase. This stretch was mainly gardens of various ages, and the grassy jungle and saplings of garden regrowth. Everything was wet from the night's rain, and my boots were soaked even before I reach the first river. The river—about 50 yards across at this point—was muddy but not in flood. Still, at this crossing it was far too deep to ford. We were ferried across in a small dugout canoe.

We made our way through grand lowland forest, occasionally broken by the course of a stream or small river. We crossed and recrossed watercourses, and in places followed the jumbled gravel bars of the Sii River on our way to the Nagore. The leeches were out in force, and I suspected the chiggers were waiting for us as well (with chiggers, the level of infestation is not evident until the day after, when the itching and swelling begin).

Mid-hike, we encountered a 6-foot-long Small-eyed Snake, a creature familiar to me from my Upper Watut camp. This individual was beautiful in a distinctly ominous way: its head and neck flattened in a threat posture, looking a bit like the cobra I had encountered in Andhra Pradesh. The porters killed the snake immediately, then told me that one of the Mirimase fellows I knew well had been killed only a month earlier after treading barefoot on a Death Adder. Snakes are neither a joke nor a curiosity to the local people in this area; they are a matter of life or death. As far as the communities in the basin were concerned, all snakes should be killed.

Our hike took about four hours. Arriving at Nagore I was greeted by the camp manager, Kurt Merg (a graduate student at the University of Florida), his colleague Chris Filardi (a graduate student at Yale), John Sengo (an honors student at the University of Papua New Guinea), and several local helpers and workmen. The greetings were warm and jovial, as the camp team had been without outside visitors for a number of weeks and were eager for the news of the rest of the world (they lived here with no radio, no satellite phone, no means of communication except a runner carrying a handwritten message—a far cry from today's communication-ridden world). I was pleased to see Kurt and Chris, as we had formed fast friendships right here in 1992 during their first field season in Nagore.

Camp was set up on the west bank of the Nagore River, which flowed cold and stony from the central range a few miles upstream. The main camp edifice was nothing more than a simple twinned set of lean-to roofs—two sloping shed frames that came together at a high ridgeline pole, upon which we laid several broad strips of plastic sheeting. This arrangement gave us a considerable expanse of rain-free work and dining area, with a cooking zone in one corner, a camp table in another, and food and cookware storage in a third corner. Nearby was a sleeping "house" for the camp staff, like a smaller version of the main house. A bit more distant was a covered tenting area where we put several sleeping tents for the researchers. These tents were shielded from the elements by plastic roofing to keep them dry and out of direct sunlight, for long-term comfort. (The strong sun and heavy and persistent rains at Nagore made open-air tenting less than ideal.) These various structures were set in small areas cleared beneath the more or less continuous canopy of the riverine rainforest. An outhouse was down a trail away from the river and the main house. It was a small bush-materials enclosure, with a plastic roof and a rudimentary slit trench in the ground.

Conditions were rustic but picturesque. It was our home in the jungle, and most everybody found it quite comfortable. I say most everybody; a couple of weeks before my arrival, a colleague of Chris's had felt compelled to depart after only a week's stay because of a severe allergic reaction to an infestation of chiggers. The man had swelled up all over and was covered in a rash that the team agreed required emergency evacuation. He was picked up by bush plane and received medical treatment outside of Papua New Guinea.

Other scary things had happened in my absence. One night a few months previous, a tree had fallen on the camp and pinned Kurt Merg and

The Nagore River at the site of our field camp. There it was a cold, stony,
clear-water stream that provided our fresh drinking water.

Michael Lucas in their tents. The large *Pometia pinnata* tree fell in such a
way that it caused injuries and mayhem but no fatalities. Kurt suffered some
broken ribs, and Michael was so frightened by this close brush with death
that he hastened to depart the camp permanently, muttering that it was under
some black magic curse.

When Kurt asked the Kunimaipa men about the frequency of forest tree
falls, they commented that *Pometia pinnata* trees were particularly suscep-
tible to snapping at their base and toppling—and that our Nagore campsite
was infested by *Pometia* that might be destined to fall. As a result, a thinning
program was conducted by the more knowledgeable arborists among the
Kunimaipa, making it easier for us all to sleep in peace.

Aside from the long-term goal of encouraging conservation of this vast forest landscape, Nagore Camp had two immediate purposes. The first was to

document the extraordinary biological diversity of this patch of forest. The second was to encourage the training of students from Papua New Guinea and the United States in the ways of fieldwork in tropical rainforest. In 1994 the trainees were Francine Wiest, Chris Filardi, and Kurt Merg (United States) and John Sengo and Robert Bino (Papua New Guinea). In addition, Chris Unkau from the environment department was visiting to observe what we were doing here in the forest: We were studying bird diversity, plant and insect diversity, seed dispersal, and seed predation; we were learning to work productively in the forest; we were learning the ways of the woods from our local informants; we were making friendships that in certain cases continue to this day.

To give a sense of a day in the field, let me summarize what each of us did on August 2, 1993. It had been cool after a "cold" night (70°F). It was overcast in the morning and despite clearing and plentiful sunshine, the day's high was only 77°F. Kurt was officer-of-the-day, but was unable to delegate work because the weekly village work team did not arrive until noon and the party was only two, instead of the requested four. Each week our camp was assigned workmen from Kakoro by the district officer, the teams being rotated to ensure that all four clans had an opportunity to make some money while we were in the basin.

Kurt operated the mist-nets and censused the avian foragers visiting a fruiting fig tree, and also conducted a portion of his seed dispersal experiment. Chris worked on one of his seed predation experiments at a species of *Terminalia;* he was monitoring the fate of fruits stacked in small piles under the parent tree. Chris surveyed foragers at the fig tree in the morning and spent the late afternoon and early evening observing bird movements from the canopy platform he had constructed in a treetop at the summit of Kautak hill. Francine spent most of her day watching avian visitors come into the fruiting fig tree. She visited the westernmost knob of Kautak hill during the midday, seeing few birds, but noting that the hilltop forest was different from that behind our camp. John collected weevils again this morning. Today's take was sixteen morphospecies, including five forms he had not previously seen. He climbed the Kautak tree platform with Chris during

the afternoon, returning well before dusk in order to prepare the camp supper. Robert, like Francine, spent most of his day at the fig tree, which he and
Francine had arranged to have under watch continuously from dawn to dusk
(bouts of watching were hourlong assignments and Robert, Francine, Kurt,
and Chris all participated on this day). Robert watched for two hours at
the tree, at dawn and dusk. He walked the trails when he was not watching,
and spent time observing foragers who visited the fruiting *Dysoxylum* tree
at Toti stream. I began my day early, with an audial bird census at one of the
tree plots. I spent the rest of the morning either working in tree plot #1 or
observing at the fig tree. In the afternoon, Matai and Papa Korau joined me
in plot #1, helping with Kunimaipa names for the trees, to be used in our
identification manual. While sitting quietly under the fig tree, I was again
visited by an adult cassowary. As on the previous day, the fully grown individual came in, foraged peacefully, and remained within view until I had to
leave for work elsewhere. Chris had two encounters with an adult cassowary.
From the pattern of ticks attached to the bird's neck, he was able to confirm
that he had seen the same bird twice. On both occasions, Chris was able to
approach to within 30 feet of the ostrichlike bird.

On a typical day, several of us would be up by 5:45 A.M. We would
quickly dress (mainly in cotton: baseball cap, khaki long trousers, cotton
long-sleeved shirt, cotton socks, jungle boots), breakfast on granola and/or a
navy biscuit and perhaps some hot tea with sugar, then march off to our first
task of the day. We would know our assigned task from checking the chart
posted daily in the main camp house. Often the first task was to open the
mist-nets. This job required hiking out to the nets, unfurling them, cleaning them of any leaves, and making them tidy and invisible. They would be
checked by someone every ninety minutes throughout the day until they
were closed. After opening the nets, I would often set off to a fixed transect,
where I would conduct a bird census mostly by ear. Other team members
might camp out in a chair beneath a fruiting tree to watch what birds came
in to feed; or sit in a blind and wait to see what animals came to feed on fruits
or seeds that had been arranged in piles on the ground under a fruiting tree;
or collect fallen fruits to measure insect predation on the seeds. This sort

of busy work (some fun, some boring, some boring but interspersed with moments of excitement) would continue throughout the morning.

Around noon most or all of us would return to camp. At this point I would immediately disrobe to escape my chigger-infested clothing and bathe vigorously in the creek to get rid of any chiggers that had made it onto my skin. I would dress in a bathing suit and sandals (rubber thongs) for the lunch break and hang the morning's clothing in the sun to burn off any chiggers hiding therein. Lunch consisted of navy biscuits with peanut butter and jelly, powdered lemonade, and perhaps a cookie or some chocolate. The meal might be followed by a brief rest on a mat in the main camp house, reading an old *New Yorker*. We would chat about the morning's events and discuss what was in store for the rest of the day.

In the early afternoon, we would all filter back out into the forest to pursue our particular interest (trees, weevils, birds, fruit, whatever). Before going out I would dress in a "fresh" set of field clothes—clothing that had not been worn that day—to reduce my chigger load. Work would continue either until a heavy rain drove us back to camp (a four o'clock rainstorm was very common) or the approach of dusk or fatigue. On the occasional afternoon without rain, the humidity would build up so ferociously that in the forest it seemed difficult to catch a breath. My clothing would be soaked through and I could think of nothing other than a long cool soak in the river. Getting out of damp clothing and into the river with a bar of soap was a luxury like no other at Nagore. I would sit on the sandy bottom and loll back in the chilly mountain water and watch imperial pigeons and mynas and Long-tailed Buzzards pass overhead.

At this point, hunger would begin to kick in, and I would plan a pre-dinner snack. The two options were popcorn and guacamole. On the best days, we made both and washed it all down with a cup of lemonade. Guacamole, you ask? The basin villages were rich with avocado trees, planted over the years. We were visited weekly by teams of women who carried out loads of vegetables to sell. The produce was carried in hand-woven string bags that are known in Pidgin as *bilums*. The women brought us sweet potato (*kaukau*), taro, yam, banana, assorted greens, and papaya (*popo*). These fresh

The Nagore field camp, our home for two-month summer research stints

vegetables and fruits were an appreciated addition to the diet of canned and dried foods brought from Port Moresby.

Dinner, cooked by one of us in accordance with the weekly schedule, would be heaped high on enameled metal plates that we had bought in a Chinese trade store in Port Moresby. A typical meal was rice and stew. The stew consisted of some canned meat (corned beef or curried chicken) mixed with sauteed onions and sweet potato, and flavored with a packet of dried-soup mix. This fairly unappetizing fare was made edible by the addition of spicy chili sauce, some curry paste, or whatever other condiments were available.

Eating this meal required some talent. The plate conducted heat very well, so it had to be handled gingerly. The hot plate would burn my lap. Consuming this steaming heap of food also undid all the benefit of the cool predinner swim. The sweat started again. It was best to dine in as few bits of clothing as possible, perhaps just swim trunks and sandals.

The highpoint of the day often was the leisure time on the riverside bench at dusk. We watched the giant flying foxes stream out of their hidden jungle colony, flapping slowly overhead on the way to their favored fruiting trees all around the basin. We also looked for Bat Hawks or other high-flying birds in the gloaming. We listened to the growing racket of the night insects as dusk turned to night. Biting insects were not a problem here at the riverside. Such was the pleasure of watching night fall in the jungle at Nagore Camp.

We ended the day by completing the camp diary (the officer-of-the-day would interview each campmate for information) and writing up field notes or carrying out some preliminary data analysis. Our field notes were the currency of our mission, so we treasured them and protected them and worried about them. If guarded and used wisely, these notes could later be converted into scientific papers or book chapters.

We were all here in Nagore Camp to document the animal and plant inhabitants of the magnificent alluvial rainforest that blanketed this old lakebed seven degrees south of the equator. In a sense we were worshiping its mind-blowing richness and diversity.

I first got a sense of its splendor from the air, flying in from Port Moresby. Let me describe our charter flight to Kakoro that June. First, our battered, thirty-five-year-old Islander aircraft rose from Jackson's Field in Port Moresby and lumbered slowly northwestward, initially over seasonally inundated grassy wetlands. Then we passed rubber plantations and logged-over hill forest, and finally flew south of the Gibraltar-like Mount Yule and the jumbled limestone pinnacles that led to Omeri Mountain. After Omeri, the gateway into the basin, we suddenly passed the last foothill. There, displayed to our eager eyes, was a vast green plain as flat as a pool table, its canopy broken only by streambeds and patches of swamp. Ahead, we gazed over the forested plain to the limit of our sight. To the left were the low Kurai hills, arrayed across the landscape as straight as a ruler. To the right rose the craggy peaks of the Chapman Range. Our plane passed directly over Nagore Camp; we could see the orange plastic roofing that marked the little encampment that Kurt and Chris and Michael had set up in April. These tiny squares of color were entirely engulfed by the huge expanse of green canopy. Crossing the basin to Kakoro in the plane took about fifteen minutes. During that time we saw no evidence of human habitation. The forests east and southeast of the Biaru River were uninhabited, forming a large natural forest reserve that local communities saw as a hunting ground. Another few minutes brought us to the Kakoro landing ground, at the verge of the foothills of the Chapman range.

Mostly what we did at Nagore Camp was scrutinize and document. Every bend in the trail brought new surprises. We encountered the giant fruits of an unnamed *Aglaia* that were larger than a softball and weighed almost 2 pounds. These dropped with a heavy thump to the ground and broke open to expose bright red arillate fruit that would be gobbled up by cassowaries. In fact, fruits of all kinds were here in profusion. The area seemed to be a vast bird- and rat-feeding station, fruits and seeds galore. We also encountered skinks and lizards, goannas and snakes. The snakes often, but not always, sat quietly, waiting for prey. The skinks and lizards rattled around in the leaves, stirring up their prey. Birdsong enlivened the forest the whole day long and mixed with the sounds of katydids and cicadas at certain times in the evening and late night. Sometimes the noise was almost deafening. When we tried to record a particular birdsong, the playback was packed with other sounds.

Most remarkable of all was that this seemingly uniform blanket of verdant forest was, in fact, a patchwork quilt of vegetation types, each patch with its several dominant tree species, mixed in with fifty or sixty that were less common. These patches were evidence of past local events that had impacted the forest (such as torrential floods, wind, fire, garden clearance, or seasonal inundation). The forest looked uniform from a distance, but up close it was a mishmash of types, impossible to characterize easily.

We found this out when doing our tree-plot mapping with the assistance of our Kunimaipa naturalists, Papa Korau and Matai. What was initially thought to be uniform "lowland alluvial forest" became a fantastically complex checkerboard of tree stands, of particular species whose identities only slowly revealed themselves to us with the assistance provided by my two naturalist informants and additional botanists from Papua New Guinea's research institutions. The giant *Aglaia* that dropped the huge cassowary-fruit (we called it by its Kunimaipa name, Toar-med), was readily identifiable by its sturdy trunk and the golden-green hue of its bark. Other trees were nondescript and much like yet other species, differing only in fruit type or in some aspect of the sap or bark texture (or even smell).

The tree life provided the base structure of the forest and was uniform and heterogeneous at the same time. It was a chaotic wonder, the outcome of history, resource availability, and competition. It was the reason the basin supported more than two hundred species of birds and some two hundred fifty species of ants. But the tree life itself was also diversity. The basin probably had more than five hundred species of trees. So diversity was built upon diversity.

We luxuriated in trying to get a sense of what was here, to document the diversity and put names on life forms that initially were not recognizable to us. Cataloguing is a uniquely human trait and it can be carried to extremes in the rainforest. Kurt, Korau, Matai, and I spent days in our tree plots, mapping and identifying every individual tree larger than 4 inches in diameter at breast height. This work was painstaking, laborious, hot, sticky, and at times boring. But it made us aware of the remarkable richness of these forests.

Over a two-year period we completed three plots of trees. The poorest had 93 tree species, the richest 178 species. Of the trees in the two abutting Nagore tree plots (sharing a common 100-yard boundary), only one third of the species were found in both plots; two thirds were unique to one or the other. Clearly, our 2.2-acre plots were too small a sample to get a handle on the vast diversity present in the forest, and the scale of the diversity and patchiness. We were merely scratching the surface.

One take-home point of our collective research work was that it was difficult to obtain samples that were representative of the basin. The high species-richness, the extreme patchiness of distribution of the plants and animals, and the rarity of many of the species gave us trouble. Rarity is a common phenomenon in hyperdiverse habitats, and it hindered us. Understanding the major patterns of abundance and distribution in the forests of

Nagore Camp, and Lakekamu Basin as a whole, will require a much larger investment than has been made—on a broader scale, with more samples. Patchiness and rarity were our two biggest surprises (we already knew the basin was rich!). Even the more common species, like the *Pometia pinnata* tree and the Common Paradise-Kingfisher, were abundant in many places but inexplicably absent in other sites within the basin. These species were seeing the environment as a set of patches, even though we tended to it as continuous lowland rainforest. The "history" I spoke of earlier must exert a strong influence on the presence and absence and abundance of species in ways we do not yet understand.

The conservation lesson we learned in the basin was a hard lesson indeed, one that took some time to sink in. It is the story of how Western scientists who get involved in nature conservation make one mistake after another when learning how to deal with local communities living in biodiverse forest lands.

Mistake No. 1 We underestimate the local people, simply because they are barefoot and wear clothing that is old and tattered. We try not to do so, but our hubris tends to win out. Our "superiority" is obvious to us: we have more money, we have access to charter planes, we have better clothes and shoes, our health and nutrition are better, we tend to talk a lot and they tend toward reticence, we must therefore know more . . .

Because the local people live in bush-material houses, inhabit hamlets without electricity or running water, and often settle disagreements through violence, we tend to forget exactly who these people really are. We tend to pigeonhole them in our subconscious. Yes, the local people seem greedy and grasping at times. But we are the only source of ready cash in the basin, and we are here only sporadically. Cash is a precious resource in an isolated region without a single road or automobile, and the locals are eager to have us part with it. In addition, we often seem so profligate with cash and goods that they think that by demanding more of us we will simply give them more, that our supply is unlimited. In actuality, our cash and profligacy are perverting and overwhelming their local economic system. Their "greed" is a product of *our* behavior and *our* presence.

In our absence, the local people share everything and operate communally. When Joe Dumoi kills a cassowary, he shares the meat with a half dozen other families in his hamlet. In societies where goods and materials are scarce, it pays to share. Joe shares today with the understanding that others will share with him next week. It is a handy alternative to buying and selling and pricing everything. Their way is best for their conditions. They are not naturally greedy. Our Western cash system produces greed.

Mistake No. 2 We tend to believe that the local people are blasé about "conserving" the rainforest. They don't seem to care in the way we do! They appear to be more concerned with their own material improvement and less with "globally significant biodiversity." Well, they are poor; our extreme material wealth makes them feel inadequate and instills in them the wish to obtain the same kind of wealth. If you think about the situation a bit, it becomes obvious that the local people have been adequately managing their wildlife and forests for a long time. If they had not, we would not be here studying it and marveling over its diversity and abundance. These local people are the *stewards* of their patrimony—the forest, waters, and wildlife that represent their only source of significant wealth on earth. All they have is the natural world they own through oral customary claim. The forest produces wood for housing, fiber for clothing, firewood for cooking, mammals and birds for eating, soil for gardening, and medicinals for healing. Their rivers and streams produce potable water, finfish, eels—even crocodiles (whose cured skins can be sold for cash). The wildlife produces feathers and fur and meat as well as song and behaviors that have imbued the local society's customs and traditions with a richness we field biologists tend to overlook. We are focused on the wildlife and not on the life of these peoples.

The local Kunimaipa people have protected and husbanded the resources of Nagore over the decades. They are the *papa graun*—the fathers of this patch of ground—who look after the land, use it, and let it regenerate itself. They grow sweet potatoes in their swidden gardens, but they also plant fruit and nut trees with the knowledge that these trees will become canopy trees in the mature rainforest and can be harvested by their grandchildren long after those who planted them are dead. Thus the forests are the local people's gardens as well, a fact that we tend to overlook. I saw this firsthand one early morning in the forest a mile or so west of our camp in Nagore. I was out in the forest, working on the tree plot, and I heard a commotion not far away—people laughing and working at something. I made my way through the undergrowth until I came upon a group of women and children, harvesting okari nuts.

This little party of harvesters was more than three hours' walk from their village, in what looked to me like trackless jungle. In actuality, these people were in their "backyard" harvesting from their informal tree orchard. The okari nut tree was probably planted in a garden a century before, and the forest had grown up around it. Ownership of the tree was passed down through several generations to one of the women who was there harvesting. This was "her" tree, and these were her okari nuts. She knew exactly where the tree was in this immense forest, and she followed a small jungle trace that I could barely discern to get here. To her and these children, the forest was friendly and welcoming and human managed and a producer of goods for the village. They saw the forest as nothing more than an adjunct to their sweet potato garden. They knew that one day this patch might be cleared again for sweet potatoes. They would leave the okari nut tree and not harm it. The garden would produce potatoes for one or two seasons and then would be fallow, quickly regenerating to saplings from the seeds in the soil. Within fifteen years it would be forest; in thirty years it would be impressive mature jungle again.

Thus the more we looked at the local people as smart, knowledgeable, and attuned to the forest, the more our work benefited from their special knowledge and advice. Only those who knew a lot in this Lakekamu jungle world would survive and prosper here. Selection was strong for people who could manage under these conditions. Our miscommunication arose because we were two separate societies, each with its own customs and interests and with little in common except the forest resources of the Lakekamu Basin.

I would get upset when old Peter, in Tekadu, would tell me point-blank that he wanted to sell a big patch of rainforest to the oil palm company, for cash. He felt this way, he told me, because he was old; before he died he was eager to experience wealth and benefit from his earthly patrimony (his forest). He knew his sons would profit one day, and he did not think it fair that he should have to do without. Wasn't Peter expressing every human's attitude about getting a break in the world? Don't we all want to be comfortable and well off or even wealthy? Why should Peter be any different? We conservationists tend to gratuitously impose values and intentions on our village counterparts. These are nothing more than twenty-first-century biases and prejudices, and they are usually dead wrong.

In interacting with Peter, I was insisting (in my head) that since he was a Kamea villager living in the Lakekamu Basin, he should be happy with his lot in life and should appreciate the same aspects of the basin that I did. This was just muddled thinking on my part. I came to the basin because it was so unlike what I knew in my home territory in the United States. But had not my people decimated North America's very best forests that held wonderful birds and mammals? Witness the remnant secondary forests of the Mississippi bottomland that now must support what may remain of the Ivory-billed Woodpecker population. Old-growth forest is gone from the Mississippi drainage. Peter, on the other hand, had taken care of his forests, but also wanted a taste of the good life: a car, a drivable road, a house with running water, a cold beer. By leasing land to oil palm operators, Peter envisioned that he could obtain enough material wealth to allow him to experience advantages that to my campers and me were commonplace back in our hometowns.

Mistake No. 3 In 1993 we naively lectured our village friends on the virtues of nature conservation. Being very polite people, they would nod their heads and agree, when in all likelihood they were wondering why we were so foolish as to cite the necessity of "conserving" forest in a place where not a single road, not a single logging camp, not a single oil palm plantation existed, and where the only human disturbance was small scale and mainly ephemeral—alluvial gold panning and swidden gardening. Forest was here in excess; it was civilization that was in short supply. I am now sure they were wondering why these foreigners thought it so important to establish a "protected area" in the basin. Protected from whom? The local people had been managing the forest for generations and it seemed today to be abundant and productive. We were bringing the proverbial coals to Newcastle.

In so many other ways we failed to communicate with one another. Let me give some examples. Raiyam and Chris Filardi were talking one day, when Chris mentioned how fortunate Raiyam was to live in this wonderful pristine environment. To that, Raiyam replied that in this "pristine environment" it was commonplace that when a small boy contracted a staph infection in a cut on his leg, the infection would grow unabated until it went systemic and threatened the boy's life. Raiyam pointed out that where Chris lived, the boy would be taken to a physician who would administer an antibiotic that would quell the staph infection before any damage was done. Here in the basin, where the health center usually lacked antibiotics because of supply difficulties, the boy would die or lose his limb for lack of a course of tetracycline or penicillin.

On another occasion, a delegation of Biaru and Kovio elders came to visit our camp, to inform me that the Nagore area actually was not the traditional land of the Kunimaipa, but in fact belonged to the Biaru and Kovio. Instead of taking the statement under advisement, I stated boldly that I knew for a "fact" that the land belonged to the Kunimaipa; my historical research in the Papua New Guinea National Archives had provided evidence of this fact. They shook their heads and vowed to prove me wrong. A week later an official letter was delivered, stating that I was working in land "officially under dispute." And this turned out to be the truth. The ownership became disputed in the late 1970s, when a gold mining company had attempted to develop a mining operation in the Sii-Nagore area. The dispute between the Kunimaipa on the one hand and the Biaru and Kovio on the other became ugly. The mining company, seeing conflict and a threat to its potential investment, pulled out. The conflict arose only because of potential economic development. Our camp on the Nagore was a new economic development, perhaps the most significant in the whole basin, and that caused the dispute over compensation to arise once again.

Here's how I got it so wrong. To begin with, I was thinking linearly and was reasoning in accordance with an oversimplified Western logic that did not necessarily apply in these conditions. For instance, my map, printed in Papua New Guinea, showed that Nagore Camp was in Central Province. Yet it was under the jurisdiction of Gulf Province. Later, when I checked with the Central Province Lands Department about the proper authority over the Nagore Camp land, I was shocked to hear the Central Province officer say, "Well, if the local land claimants are living in Gulf Province, then that land must properly be under the authority of Gulf Province—official map be damned!"

Furthermore, I had looked at the diaries of patrol officers and had satisfied myself that at first contact with Western authorities, the Nagore area was inhabited by the Kunimaipa. Actually, in precontact days the lowland areas like Nagore were essentially uninhabited and uninhabitable because of disease and fear of attack by enemy clans. To avoid surprise attacks, the traditional peoples situated their villages on ridge-crests of the adjacent hill forest, not in the flat lowlands. People did not want to live down in the low-lands where conditions were ripe for malaria and other diseases, and where garden land was rather poor compared to the hill forests. The vast lowland basin was a large hunting reserve that was never settled until the German gold miners taught the local inhabitants about the value of gold, and until the Australian patrol officers sought to settle people in concentrated communities where they could be censused, treated for health conditions, and regulated. This resettlement reduced travel strain on the patrol officers, but it also brought together ethnic groups that were traditional enemies.

Kakoro village, which I came to call the snake pit for its intrigue and petty provincial politics, is a monument to the mistakes made by white people (including me) in Papua New Guinea. A patrol officer named Brown "created" Kakoro by fiat, because of its proximity to the flat land for an airfield, and because it seemed central to the scattered villages that were around the verges of the basin. This brought the Kamea down from the north, the Biaru from the northeast, the Kovio from the south and southeast, and the Kunimapia from the east and northeast. The Kamea (also known as Kukukuku)

were excellent fighters, feared by all and left alone. Their heartland was in the hills to the north and northwest (the Watut, Anga, Kapau lineages). The Biaru originated in several highland valleys in Morobe Province, up the Biaru River. They arrived in the Lakekamu Basin in the 1970s in search of gold and settled in Kakoro because it was a district station. The Kovio were the true lowland people (sometimes called bush Mekeo) with relationships to the coastal Papuan people. The Kunimaipa were of Goilala stock, with roots in the hills and mountain country to the southeast in Central Province.

The fearsome Kamea tended to steer clear of the politics of Kakoro. The Biaru developed a political alliance with the Kovio (a case of the newest and weakest group allying itself with the most politically connected group). The Kunimaipa were opposed by the Biaru-Kovio alliance. In 1993 the district officer in Kakoro was a Kovio man, so the balance tipped strongly in favor of the anti-Kunimaipa alliance. Through a historical accident, I was allied with the Kunimaipa, which turned out to be a source of considerable trouble for me and my fledgling field program.

The historical accident was that in 1982, when I came with my new wife to spend my honeymoon in the forest southeast of Kakoro, I was met at the landing ground by a group of Kunimaipa who claimed to be the landowners of the area where I wished to camp. I worked with them happily for two weeks in the forest and never forgot how well that went. So when I returned years later, I believed in my heart that the Kunimaipa were the true "landowners" of the forest southeast of Kakoro. I engaged with the Kunimaipa in 1992, and they selected Nagore as the best area for researching wildlife. They were spot on; the site was superb, the best I have ever worked. How would they have known that if they had not been the landowners?

Well, Nagore and the area around the Sii River were officially under dispute because of the claims of the Biaru and Kovio. The controversy would simmer at low heat for years until an outsider appeared with a desire to carry out some form of development there. First it was a gold mining operation, and now it was a biodiversity miner (me). Yes, to the combatants in this argument I was seen as nothing more than a miner who might produce revenue for whichever group could "prove" ownership of the resource.

Ownership—there was the problem. When the only form of proof is stories of ancestors, poorly marked gravesites, and planted fruit trees, resolving land disputes is difficult indeed! A land court in Kerema would hear both sides and make a decision. Then, not uncommonly, the decision would be reversed on appeal. It was very much "he said" versus "she said." Not a satisfactory way to resolve a dispute about the most important resource of the Papua New Guinea rural people.

My big mistake was siding with the "true" owners. I was right in the Western sense (my research showed that the Kunimaipa had villages in the Nagore area in the precontact period), but that made no difference. My long-term goal was nature conservation, which could only be achieved if there was consensus in favor of this activity among the decisionmakers in Kakoro and Kerema. In my haste to "pick a winner," I had guaranteed failure. It was not a winner-take-all situation. These various ethnic groups shared the Lakekamu Basin and shared Kakoro; they had to learn to get along even though their past was replete with strife and actual combat with spears, bows, and arrows. I was using a critical and analytical mindset to make decisions on right and wrong, when in fact I needed to use a more anthropological mindset to look for common ground and consensus.

Conservation International is still working in the Lakekamu Basin, but I have not shown my face there since 1997. It would be a mistake for me to return, because my foreign ideas and white face would only bring reminders of conflict and dispute. Instead, Papua New Guineans Philip Lahui and Kepslok Gumilgo are patiently working through local-level government to create a consensus among all four ethnic groups and to move ahead with a very incremental plan to achieve conservation.

Here is the new thinking on a way forward: Philip first obtained a Provincial Executive Council decision that supported developing a large-scale resource-use plan for the entire catchment area. This plan is being followed by a government-led process that will attempt to map the future of the basin. It will explicitly include both economic development and nature conservation. The group will lay out gardening areas, mining areas, possible plantation zones, all embedded in a large matrix of conservation (traditional hunting) lands. In all likelihood, this plan will be certified by the government and then a formal "conservation incentive agreement" will be negotiated between Conservation International and the four local-level governments. The agreement will offer annual development support in exchange for following the management plan that has been certified by the government.

The agreement is, in essence, a purchase of the globally significant "environmental existence values"—what makes Lakekamu special to the outside world. These environmental services (water, intact forests, and the like) and the intact biotic assemblages (the native plants and animals) are of global significance, and their maintenance has a market value for which conservation organizations are willing to pay. If our plan works well, it will lead to the creation of a national conservation area through PNG's Conservation Areas Act. Concession payments will offset the cost of managing the conservation area and will provide some leverage for localized economic development in a part of Papua New Guinea that, because of its physical isolation, has few viable economic opportunities.

What will the Lakekamu Basin look like in twenty-five years? If our work goes well and Philip and Kepslok succeed in their negotiating and planning and mapping, then the basin may become a remarkable place, where an ecological field station attracts research scientists; a high-end ecotourism lodge draws specialist visitors who would like to see a cassowary and a crowned pigeon and a Vulturine Parrot at close range; and local communities direct these operations and also manage small-scale alluvial mining and a thriving betel nut plantation system that together generate sufficient income to pay for the necessities of life. The bulk of the basin would remain cloaked in rainforest as it is today, and the vast uplands to the east would be a protected watershed that rises into the mists. The natural wonders of the Lakekamu Basin would remain for the world to wonder at.

CHAPTER EIGHT

Pitiful Scraps of Forest

Cebu has long been noted for its deforestation.

N. J. Collar,

"Extinction by Assumption; or,

the Romeo Error on Cebu," *Oryx* (1998)

WHILE WORKING AT COUNTERPART International, I spent a fair amount of time on the road examining rural agricultural systems, and I found merit in the polyculture aspect of *Forest Gardens* (Chapter 6), though I did not think it could take the place of monocultures of certain staple crops (maize, wheat, rice, for instance). Nonetheless, in countries such as Madagascar, Sri Lanka, and Philippines, too much hilly upland forest was being devegetated in order to plant row crops for cash. Soil erosion was a big problem, as was scarcity of fuelwood, poor home nutrition, and drinking water contamination from fertilizers and pesticides. I believed that *Forest Gardens* could help address this suite of issues facing poor rural agriculturalists.

In this chapter I examine Counterpart's role in introducing the *Forest Gardens* system to Cebu Island in the Philippines. The Philippines is a rainforest nation whose forests have been decimated, in stark contrast to the situation in Papua New Guinea and other forest-dominant tropical nations. Cebu is of particular interest because of its history of extreme deforestation and the impact of widespread habitat conversion on a tiny, beautiful, imperiled endemic bird—the Cebu Flowerpecker. Recent efforts have been made to conserve the flowerpecker and the last remnants of the native forest. Cebu is an island that exemplifies the conflict between human development and nature.

❧

The Philippines, lying on the western edge of the South Pacific, five hours air flight northwest of Papua New Guinea, is unlike any other nation in the South Pacific or Southeast Asia. It is populous like Indonesia, but Christian, not Muslim. It is industrialized like Malaysia, but not nearly so regimented. It differs from nearby Vietnam because of its American rather than French colonial history. Its differences with Papua New Guinea are perhaps greater than with any other neighbor. PNG is forest rich, the Philippines is forest poor. PNG is very Australianized, the Philippines very Americanized. PNG supports a remarkable wealth of Christian sects, whereas the Philippines is predominantly Roman Catholic—but with a Muslim area far to the south. The Philippines is strongly Asian, PNG strongly Oceanic. The Philippines supports huge urban centers, PNG is largely rural. The Philippines' main

Preceding page: The Cebu Flowerpecker

staples are rice and corn, those of PNG are primarily sweet potato and taro. The list goes on. Perhaps one of the few shared habits of the two nations is the consumption of pork in quantities. Thus, this story is about a tropical rainforest the polar opposite to that of Papua New Guinea.

I first traveled to the Philippines in November 1997, with David Vosseler, my boss at Counterpart. We took a long overnight trans-Pacific flight to Manila International Airport. The international terminal was perfectly adequate, but David and I had to connect with a domestic flight that would take us to our destination on Cebu Island, in the south of the country. To accomplish this, we were told, our only option was to rent a car and driver, then venture into the swirling chaos of urban Manila traffic and circumnavigate the airport grounds to get to the dumpy little domestic terminal on the other side of the airfield. Instead of taking a taxi or internal bus service or monorail (as one would do in a score of other airports), we had to go through the insane process of renting a car (and a driver), and then, upon seeing the traffic on the streets of downtown Manila, praying for salvation. The crush of cars, buses, trucks, and jeepnies, all pushing into and out of lanes and tooting their horns, was an awful spectacle. We were jetlagged and in a huge hurry to catch our flight to Cebu. We encountered smog, swarms of urban poor hawking their wares from street corners, noise, and way too much traffic. Our Philippines-total-immersion course was starting in our first hour. It seemed as if we would never make it to the domestic terminal.

By a small miracle (a delayed flight) we made it to Cebu Island, carrying six large boxes of computer equipment for our new Counterpart office in Cebu City. The relatively short flight was uneventful, and the view from on high provided tantalizing glimpses of cloud-bedecked volcanic cones, volcanic lakes, wonderful natural harbors, and lots of islands and shimmering tropical sea. The Philippines is nothing less than a fractured effusion of islands along the Pacific's "Rim of Fire," with extinct and active volcanoes scattered the length of the archipelago.

At the Cebu City airport we were met by the recently appointed project director, Leonardo Chiu, and his team, who drove us to the city. Here was another test of stamina. The airport on Mactan Island was linked to

downtown Cebu City by a series of narrow streets that zigged and zagged
through a periurban wasteland of warehouses, big-box shopping malls,
neighborhoods, roadside stalls and shops, creeks now serving as sewers,
and burgeoning traffic. The drive took about an hour, but it seemed like much
more because of the heat, glare, and ugliness of it all. Here was unzoned and
unplanned development at its worst—a concatenation of former villages and
little neighborhoods all cheek by jowl, merging amoeba-like into one vast
low-rise tropical city of shacks, offices, shops, and factories. Most distressing
was the fact that this land had once been a rainforest bordering on a diverse
coral reef ecosystem. Now it was the worst of the urban tropics.

Cebu is the place where the Spanish navigator Ferdinand Magellan
met his end in the year 1521, on his fateful round-the-world cruise. He was
killed in battle by Cebu's renowned historical figure, the local tribal leader
Lapu-Lapu. We drove by an imposing statue of Lapu-Lapu near the site of
the bloodshed on Mactan Island. Cebu itself is a 150-mile-long sliver, sand-
wiched between the islands of Negros and Bohol in the Visayan group of
the southern Philippines. Cebu City is a commercial hub of the Visayas and
one of the Philippines major business centers, though probably little known
to most readers. Cebu Island is mountainous, tropical, and infamous for the
destruction of its forests and the extinction or near-extinction of its endemic
plant and animal species. Cebu was home to Spain's earliest settlement in
the Philippines and once supported immense stands of tropical timber that
fed Spain's ship-building industry.

Downtown Cebu City is a pretty fair place, as cities go. It is rather like some
sections of Jakarta, with nice buildings, restaurants, shopping, and a central
traffic circle with featured venues on it. David and I were put up in the Diplo-
mat Hotel, a two-star institution with the basic amenities. The rooms had a
musty smell (typical), the bathrooms were cramped and uncomfortable, the
air-conditioning was only semifunctioning, and the noise from the street at
night kept us awake. But we were okay. This was life in urban Southeast Asia.

Our Counterpart office was not far from our hotel in the city center,

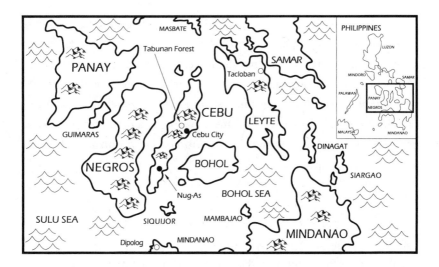

on a little side street in a low apartment building that had been converted to office space (a common phenomenon in Southeast Asia). Here, David and I met over the next several days with Leonardo Chiu (nicknamed Nards) and his team of *Forest Gardeners.* We were launching a five-year-long, U.S. government-funded agricultural restoration initiative.

Regardless of the merits, turning a concept drafted in Washington into a successful field environmental program that improves the lives of the rural poor in Cebu is no small challenge. Many steps lie between the initial flush of receiving a successful grant letter to the triumph of changed lives. We needed a strong support team back in Washington, and we had to build, from scratch in this case, a field implementation unit in the Philippines. It required a director, office space, office support staff, management and administrative systems, dedicated field staff, and, most important, one or more field sites where the impoverished rural populace was willing to subject itself to the potential for a change in behavior based on these new ideas. To succeed, our project would require a lot of time, the hard work of a well-trained staff, and some measure of luck.

Regardless of the difficulties, we were moving ahead at full speed. Leonardo was forceful and action oriented and outspoken. We already had staff, several field sites, and a program to introduce. Uncertainties of culture and communication remained, however. David and I spoke to Leonardo about the details of the plan, but did he really hear what we were saying? When Leonardo made complex pronouncements on the simmering politics of the rural poor, did we really understand what he was telling us? Many issues had to be taken on faith, it seemed. We forged a partnership based on optimism, a common procedure in the world of international environment and development projects.

Our first pilot *Forest Gardens* site was about 80 miles south of Cebu City, in the Barangay of Nug-As (Nug-As District) in the uplands of Alcoy municipality. After a couple of days of discussions and planning sessions in the city, we piled into two rented jeeps and drove down to Nug-As to have a look at the situation in this upland community with its curious name.

We drove along the east coast from the city down toward the southern end of this narrow and hilly island. The "highway," the only artery out of the city leading south, was nothing but a city street lined with shops. The single southbound lane was choked with buses, trucks, jeeps, and cars. Some strategically located road construction required detours through suburban neighborhoods, which explained our three-hour slog to Nug-As. The first hour was the worst; our progress gradually improved, and the scenery became more and more interesting as we entered the rural south of the island.

The east coast of Cebu is very pretty, with beaches, tiny beach communities, and lots of greenery. Abundant mango trees, flower and vegetable gardens, and yard plantings make each community picturesque. Almost anything can grow in such tropical conditions, and the rural people had a real flare for ornamental horticulture. In addition, nearly every community seemed to have banners announcing a carnival or an anniversary or some such. I was reminded of rural Indonesia—the scenery was verdant, bucolic, and pleasant to the eye. In one beachside community we stopped to examine an ancient Catholic church of Spanish ancestry. Parts of southeastern Cebu were the quintessential South Seas paradise.

At Alcoy we turned inland and crossed some scrubby savanna lands, then climbed steeply over a ridge that was a mix of scrub and old fields. It was obvious that this area suffered significant dry seasons. The drought-prone situation was exacerbated by a soil that was dominated by porous limestone rock. Uplifted coral reef capped the hill country. Things were crispy here in the late November dry season, as the rains had not yet begun. We drove through deep green glens on our way across the top of the ridge, then descended into the remarkable hidden valley that was Nug-As. It was a world apart, a small agricultural basin at 3,000 feet elevation, surrounded by hills on all sides, a little Shangri-la not unlike the valley where I had camped in Arunachal Pradesh, but quite a bit less isolated and quite a bit more affected by human activity.

Life in Nug-As centered on its marketplace. To see it was to understand something about rural upland life in the southern Philippines. The stalls were filled to the brim with vegetables and fruits and fried foods and colorful clothing and all sorts of oddments. Deep-fried chicken feet were a featured delicacy. People of all ages surged through the narrow passageways of the covered market in search of bargains. Agriculture was obviously the driving force. On this day the sun shone brightly in the dark blue sky, and huge white clouds billowed up in the east. Clots of people talked and laughed and smiled at the sound of loud music played through large metal speakers. Men lounged, smoking cigarettes. Women perched over their goods shouting to passing customers. Life blazed brightly.

The people of Nug-As farmed corn (maize) as a staple crop, as well as an array of vegetables for the market. These were complemented by tree crops (guava, citrus, jackfruit, banana), poultry, and pigs. Rattan was harvested from the forest remnants. And of course just about anything of cash or subsistence value was collected adventitiously to make ends meet in this impoverished corner of Cebu.

Our *Forest Gardens* project planned to build on the successes of a just-completed German aid project that for several years had fostered market gardening to help build a cash economy in Nug-As. The fertile bottoms of the valleys grew mainly corn, and the hills were tended for pole beans, root crops, and the like, which were trucked in large part to Cebu City to sell to the restaurants and hotels. The German project had done little on the environmental front, encouraging the community to invest in row-crop agriculture that depended heavily on fertilizer and pesticides. This was the "old" market agriculture that we hoped to reform. How much we could accomplish over the five-year tenure of our USAID grant was difficult to fathom at this point. Could we instill an environmental ethic in a community that had focused on maximizing cash flow and profit from city-focused market cropping?

Here in Nug-As, our *Forest Gardens* team was going to keep the project fairly simple and not try to reinvent agriculture. We did not want to put these farmers at risk just to test all the intricacies of a complex experimental system. Our program would focus on encouraging polyculture cropping, mixing crops and using less pesticide and fertilizer, and greening the agricultural system wherever possible. We had built on the existing cooperative system, in which farmers teamed up to achieve economies of scale and to share profit and risk. The rural Philippines is highly political, and cooperatives are the true backbone of the rural political system.

Let me make an important point, one that is often overlooked in the United States. To better our environment, it is most efficient to combine governance, developmental planning, and environmental best practices in a single decisionmaking system. The cooperative can achieve this integration. In the United States we tend to keep local politics, enterprise, and the environment as stand-alone issues that are argued on their individual merits. Yet these threads can become strong only if woven together. Imagine if, in the United States, our local, state, and national governments actually planned development in consensus fashion. They could achieve an integration of enterprise, social, and environmental objectives, looking at long-term economic costs and recognizing the noncash values imparted by the various services provided by the natural environment (fresh water, clean air, and fertile soil).

Our Nug-As project focused, then, on greening up the production of market vegetables, intercropping with valuable shade trees, orchid horticulture, and soil improvement; increasing value and attempting to decrease environmental degradation; building up soil; stabilizing hillsides with tree planting; providing small loans for the creation of local businesses to build the local economy and make it less dependent on outside interests. Certainly this was not the Analog Forestry envisioned by its inventor, Ranil Senanayake, but it borrowed some of Analog Forestry's best practices. We were taking baby steps, because change in poor rural communities is difficult and risky.

Our team had to battle against a number of ingrained issues that plagued the farmer working in the rural Philippines. Chief among these was land tenure. Virtually all of the agricultural land in the Nug-As Valley was owned by absentee landlords, who leased the land to the farming poor who lived there. These valley farmers were in essence sharecroppers, who had to make a living by farming for a landlord whose only interest was income and who had little concern for the welfare of the farmers. Life was precarious, what with necessary payments to the landlord and input costs for seed and fertilizer. Other troubles arose. El Niño droughts could not be predicted, and long dry seasons were potentially devastating. Lifestyle issues involving alcohol, cigarettes, and gambling (especially cockfighting) made it difficult for a family to get by in the countryside of the southern Philippines.

Cebu was a typical Southeast Asian island. Its weather was dominated by annual patterns of winds that brought the wet and dry seasons to its land. In the dry season, water was in such short supply that tanker trucks hauled drinking water up the hill to Nug-As to sell to the residents. Having to pay for water and for the right to farm land annually was a double whammy for these residents.

The land here was harsh. All but the very best bottomland fields were little more than fields of jagged coral rock, between which were interspersed bits of soil. Farming was difficult to carry out and not very productive. The lack of forest cover had long ago led to massive erosion on all the hilly slopes that carried away the once-rich soils. Back in the United States, we tend to forget about issues of soil and water. In places like Nug-As and a hundred thousand other communities like it, issues of adequate soil and sufficient water become fundamental, a matter of the good life versus a marginal existence.

Planning the installation of experimental *Forest Gardens* in the Nug-As Valley
was very much a social process that involved the community.

In America, when we speak of the environment, most of us think of
the Endangered Species Act, Al Gore, and the Kyoto Protocol. The politics
of polarization have highlighted certain divisive environmental issues to use
as weapons for winning elections. Those wedge issues tend to be the less
meaningful ones. In fact, the most important environmental issues are those
that are simple and straightforward: issues of soil and water for crops, water
for drinking. These are issues in the United States, but they have been hidden
by battles over more divisive, less significant issues. As a result, we will not be
facing up to these really fundamental issues for another couple of decades,
at which time solutions will be far more difficult and expensive. Such is the
case in other countries as well. Civil society is often weak about facing up to
the long-term environmental issues that incrementally grow in importance,
year by year. The easy way forward for decisionmakers and politicians is to
put off the difficult decisions.

The Cebu farmer is caught in the clash between economic imperatives and long-term well-being of the land. Sharecropping forces the farmer to think of making annual payments. The economics of debt and interest payments also force the farmer to focus on the short term: make the land pay today. By contrast, protection of the natural resource base (forest, soil, aquifers) is a long-term issue that requires planning and forbearance, and working for a better future. Some fields may need to be retired for fallow periods, to allow the soil to recover. Or we may require plantings that do not produce in the first year or two (most woody tree crops). If a farmer needs to generate pesos for his landlord, he may elect to raise a crop of carrots rather than a crop of shaded cardamom. Over the years, the cardamom would produce more income, but because of economic pressures the farmer can only plan in one-season increments—thus the unfortunate interplay between the discount rate and the environment.

Life in Nug-As revolves around cockfighting (*sabong* in the Philippines). Most of us recognize cockfighting as a sport distant from our home culture, something we may have read about in the *National Geographic,* perhaps. It is illegal in all but a few of our United States. We know it is colorful, bloody, perhaps cruel, and tightly linked to gambling with cash by the rural poor. We also probably know it is a man's sport; its closest analog in U.S. cities would be dog fighting or horse racing. For the uninitiated, I can provide some basics drawn on what I saw in Nug-As.

Male chickens (cocks or roosters) are lovingly fed and handled and encouraged to fight in home practice bouts. Those with the most aggressive nature become favored gamecocks to be entered into formal competitions. Cockfighting builds on the pugnacious dispositions of these birds, a product of the species' social behavior. In nature, cocks of the Red Jungle Fowl (native to India and Southeast Asia) are polygamous and manage a harem of females. Each cock keeps his harem by driving off any other cocks who attempt to approach his females. The cock thus has mating privileges and thereby gains "fitness" (more of his genes are transmitted to the next generation).

Cocks have a spur on the back of their lower leg, which is used to spear and cut opponents. In nature, cocks fight by lashing out at each other with their spur, seeking to do damage to the rival. Cockfighting in human society simply formalizes the natural propensity of cocks to fight. The ante is raised by adding a manmade spur to the cock's leg, a small cutting blade that the successful cock can use to brutal effect. It can be a fight to the death.

In places like Nug-As, cockfighting is an all-consuming sport to the local men, who groom their birds for fighting and who bet their hard-earned cash on the fights. Cockfighting arenas (game pits) are constructed and become the Philippine version of a bullring, with a crowd of cheering, betting, excited men watching the contact between two chickens, each seeking to destroy the other. Huge amounts of money change hands in the excitement, and when the blood is up, many of the poor borrow money at high rates to continue their betting, victims of predatory moneylenders who prowl the verges of the game pit.

A natural additive to this wondrous entertainment is *tuba* (pronounced *too-BAH*), the locally brewed coconut palm wine that is drunk in the Philippine countryside. Dreadful stuff, it is the elixir of life in places like Nug-As, where little comforts like tuba and a Saturday cockfight can make a rather grim life bearable.

In the patriarchal society of the rural Philippines, the woman's lot is probably not a fine one. Suffice it to say, the female probably rarely gets much of the tuba and usually misses out on the cockfight, as there are chores to be done (cooking, washing, weeding, nursing, firewood collection). Improving the lot of the rural poor in the southern Philippines will require addressing an abundance of gender issues.

Over the first three years of the program we enlisted forty Nug-As farming families to implement *Forest Gardens* principles in their fields. They were planting permanent windbreaks along all field edges. They were mulching. They were planting fruit trees in selected sites where fields intersected. They were reducing their dependence on pesticides. People's lives were improving, but what of the natural environment?

❦

One of the objectives of the *Forest Gardens* system is to promote biodiversity conservation. The system seeks to improve agriculture, improve lives, *and* improve the environment—to make a place for nature in the human-managed ecosystem. In order for our project to document improvements over time in the Nug-As Valley's biodiversity, we had to make a baseline survey of what lived there at the outset. The team surveyed the local residents, finding out what wildlife they encountered in their daily rounds of the fields and remnant forests. The process gave us a solid initial base of knowledge.

In addition, I led a field team to survey the forest patches that Nug-As still supported in the less accessible corners of the valley. Ten of us (four from Counterpart, six from the village) spent four days in the rocky hills south of the village, in search of wildlife. It was a sobering contrast to what I had encountered on similar surveys in India and New Guinea. The largest patch of forest in the Nug-As highlands was probably no more than 300 acres, all of it perched on the most rugged karst terrain. The floor of the forest had no soil. Instead, we found jumbled mats of old coral reef, interspersed with sinkholes. The "forest" was forced to grow on this substrate. Needless to say, the forest was a dwarf variety, in most places no more than 30 feet high, stunted and gnarled. There was a closed canopy, but all of this woody vegetation was secondary. I would guess that the whole upland area had been cleared and cropped, which led to complete erosion of the original soil and uncovered the coral rock. Because of the loss of soil, the fields were discarded and allowed to regenerate slowly into this dwarf forest. I have never seen a more paltry forest in the tropics.

That said, the forest was the richest habitat we encountered in the area. It supported a decent population of the endemic Cebu Black Shama, a threatened thrush that lurked in the wettest, darkest glens and sang its mournful song in the early morning and at dusk. At one point in the late 1950s, scientists reported that this endemic forest-loving shama was nearly extinct, but apparently this was a misapprehension brought about by some undertrained survey teams unable to identify birds by their song. The forest was also home to the Elegant Tit and about a dozen other slightly less interesting species.

On our second day in the forest, I encountered a tiny male *Dicaeum* flowerpecker. From my Philippine bird book I identified it as *Dicaeum quadricolor,* the mythical Cebu Flowerpecker. Thought to be long extinct, it had been rediscovered by a young British ornithologist only four years before, while he was conducting surveys of the Tabunan Forest about 80 miles north of Nug-As. At the time of his discovery, the Cebu Flowerpecker was considered one of the rarest birds on earth (with fewer than a dozen living individuals).

Was it possible that I had located a second population of this perilously threatened species? Was it possible that this little patch of upland forest held one of the earth's rarest species? I worked to get a better look at the bird I had seen imperfectly. In my first observation, I had seen only the front of the bird, but had not seen the lovely, distinctive patches of bright color enlivening its dark blue back.

Returning to the United States, I sent emails to colleagues reporting that the Cebu Flowerpecker might inhabit the remnant forests of Nug-As. This report was met by skepticism on the part of one Philippines bird expert, but others were interested in the possibility. Within a year of my report, a British team had surveyed Nug-As again and indeed confirmed the presence of this rare species in Nug-As and at nearby Mount Lantoy. All of a sudden, little Nug-As was on the map as home to one of the world's most endangered birds.

Otherwise, our survey at Nug-As was not terribly heartening. The forest we examined was heavily cropped for timber, rattan, bamboo, and other forest products. It was not surprising, given the large human population in the valley and the poverty of the community. One would expect heavy subsistence pressure on the forest that remained in the area, no matter how far from the village.

Having had a glimpse of one of the world's rarest birds, I wanted to have a better look. After the Nug-As expedition, I arranged to drive up to the Tabunan Forest in Central Cebu National Park. We drove up into the mountains behind Cebu City. These critical watershed hills were much overdeveloped and overcleared and oversettled—too many people and too few controls over the use of hilly lands that served as the main water source for the big city.

After a little more than an hour of driving, we came to the boundary of the national park. I saw the park sign, but no sign of natural parkland. Everywhere I looked I saw steep cultivated hills, bare of forest and infested with row crops of cabbage and carrots and other market crops. Where was the forest? Where was the protected land? I was told that because of the land tenure crisis on Cebu, the only "free" land for landless migrants was in the national protected areas. All the other land was held by greedy landlords, who prevented squatters from settling on the private land. As a result, Central Cebu National Park was a vast squatter community of landless migrants. And productive it was, indeed. Nearly every scrap of land was gardened. The officials of the Department of Environment and Natural Resources were not doing their job to protect the park from destruction.

We drove to the village of Tabunan, a hamlet deep in the heart of the park. Having villages persist in parks is probably not a way to keep forest lands protected. I was not much impressed with this community. It had a very urban feel—radios blaring rock music, people tinkering with motorbikes, little planting around the houses. However, a nature guide named Oking appeared and offered to take me up to see the rare little bird.

From the village I looked around, and frankly I could not see the famous Tabunan Forest, home of the Cebu Flowerpecker. Where could a great forest hide, after all? I followed Oking up a steep, muddy trail through pasture land and gardens. As we walked up the rain-slicked trail, I could finally see some remnant forest over the crest of the hill. The area was extremely rocky, but there was actual ground to walk on (unlike the Nug-As Forest). Close to the forest, the karst took over and we had to clamber from coral boulder to coral boulder. In some places we had little ladders to get us over huge rocks, and in other places we had to hop from rock to rock to get over deep, dark fissures. We were now in the "forest," with big canopy trees like the ones that used to blanket this entire island.

Before long we were on a little bamboo platform high above the ground.

This was Oking's flowerpecker observation platform. He told me that it was ground zero for the flowerpecker. If we waited and watched, the bird would appear. Within a half hour, Oking raised his hand and told me he could hear the high notes of a male Cebu Flowerpecker nearby. Another ten minutes passed and then I had the colorful little songbird in my binoculars—tiny, sprightly, gray, blue, red, and orange—perfection.

The Cebu Flowerpecker is one of those creatures alive and yet perhaps already functionally in the realm of the extinct. To think of this sublimely beautiful little songbird teetering on the brink of oblivion makes my heart ache. Its simple but absolute beauty and vivacity makes such an impending eventuality all the more unfair. These tiny birds, only 3½ inches long, are filled with life and song and color—how can humankind be so oblivious and uncaring?

Looking around, I saw that the bird's livelihood depended on a tiny scrap of old-growth forest, no more than 300 acres total. Signs of clearing and timbering were everywhere. I could hear chopping not too far off in one direction. I heard the report of a shotgun in another direction. Tabunan Forest was endangered, as was the little flowerpecker whose life depended on it. It was an extremely sobering discovery. Perhaps other remnant forest patches like the one I visited in Nug-As could serve as backup reservoirs of habitat for such a wee bird. (I assume a flowerpecker as small as this one could survive in secondary forest.) Subsequent field studies have located the flowerpecker in three forest remnants on Cebu. I wish I could call this good news, but given the lack of protection of these patches, and the growing population hungry for timber and firewood, I think extinction of the flowerpecker and many lesser-known forest creatures endemic to Cebu may be merely a matter of time.

❦❦

The vast unbroken forests of New Guinea's Lakekamu Basin are, in a way, a window to the past. A century ago, Tabunan's forests were thick and extensive and untouched like those of Lakekamu. To flip the scene, might it be that the pitiful scraps of forest that today constitute the Tabunan Forest are a glimpse of what the forests will look like in Lakekamu a century hence? Is this the inevitable fate of the entire tropical world's old-growth forests? The creeping effects of population, swidden agriculture, landless migration, and demand for firewood and lumber can be overpowering, as we can see on the island of Cebu, which a few hundred years ago was famous for its majestic forests.

Can the conservation interests of the world stop the onslaught on the last of the world's grand tropical forests? Authoritative studies indicate that protected areas are one piece of the puzzle; but as we see from Cebu, not all parks can work if they have the wrong social and cultural imperatives (landless migration combined with impotent government oversight, in the case of Cebu). Parks are certainly not the only answer. Given huge increases in rural populations, cries of "land" and "poverty" will lead to the conversion of many forests that are currently under protection.

No question, the single greatest threat to the world's rainforests is population: people who need land for homes and gardens, people whose desire for houses creates global demand for lumber, people who consume more than their share. If the world cannot manage its population in a rational way through smart family planning and strong economic incentives, problems galore are ahead for the earth's remaining rainforests.

More institutions need to get involved in population issues. More NGOs need to go to war against the government agencies that bend to the will of vocal minorities who dislike family planning for religious reasons. More people need to become responsible stewards of the natural world. For we all know that the natural world is the only world we have, and that all that is valuable in the developed world owes its existence to the natural world and to the many bountiful services that it provides us, free of charge.

The world needs to fix the agricultural mess it is in and address perverse government subsidies, overproduction of certain crops, and the food underproduction of many nations. Even the U.S. "food programs," which sound meaningful in the abstract, do harm to the agricultural sectors of developing countries by swamping local markets with shiploads of cheap (or free) produce that has been stockpiled by the U.S. government.

The world's agriculture is in trouble: look at the massive erosion of topsoil; agriculture in drylands that cannot support crops during drought years; divisive politics that cause the wholesale movement of peoples that, in turn, creates famines and food crises.

Here are some possible first steps toward a solution: (1) halt price supports for "protected" agricultural sectors; (2) establish government policies that reduce use of pesticides and encourage greener systems of husbandry; (3) foster permaculture and tree-crop plantings in high-relief, erodable lands; (4) cease grazing of drylands around the world, especially in northern Africa, central Asia, and the western United States; and (5) encourage site-stable agricultural systems in areas where human populations are escalating.

The future of agriculture is tightly linked to the future of the earth's tropical forests. This linkage must be better understood and managed, to make agriculture more efficient and green, and to preserve as much of the remaining forest as possible. To succeed we probably need to shift to an economic accounting of national wealth based on stasis rather than growth, for the growth model will inevitably lead to the destruction of all of the world's rainforests.

There is something terribly poignant about watching that lovely flowerpecker sing its heart out in the canopy of the Tabunan Forest. The plight of this species exemplifies all the frailties of an island environment under the press of a society in which the human population expands and governance is weak. The endangered Tabunan Forest lies in the heart of a national park, yet the park is unprotected and uncontrolled, its resources being stripped in broad daylight. This upland area stands at the top of the catchment for Cebu City. Because of uncontrolled devegetation of these uplands, the city is parched during the dry season and flooded during the wet season. City leaders make proclamations about catchment protection and revegetation, yet implementation of such declarations is lacking, as is political will. Settlers continue to move into the uplands in search of land and opportunity. The future of the flowerpecker *and* the city at the foot of the mountains remains in doubt. Cebu Island serves as the poster child for misguided development, in which environment, land tenure, and politics conspire to hinder the best intentions of principled people.

CHAPTER NINE

Lemurs, Vangas, Chameleons, and Poverty

The wooded area was of the humid forest type,

with large trees heavy with lianes and mosses.

Tree-ferns were common . . . It was cool here,

with mist lying over the lower ground in the morning,

the hills rising through it like islands

through a snow-covered frozen sea.

Austin L. Rand,

"The Distribution and Habits of Madagascar Birds"

Bulletin of the American Museum of Natural History

(1936)

MY JOB AT COUNTERPART International sent me to far-flung lands. In 1998 the U.S. Agency for International Development (USAID) announced an environmental funding initiative in Madagascar. The "request for proposals" was a call for environmental organizations to submit funding proposals to do conservation work in Madagascar. This initiative caught the eye of Counterpart's leadership as a possible new source of institutional business. I was nominated to scope out environmental project opportunities in Madagascar. I booked my airline tickets, bought a *Lonely Planet* guide, a bird guide, and chatted with my friends at Conservation International about where to go and what to see while there. My planning, although minimal, did include some help from a travel agency that specialized in backcountry travel in Madagascar. Travel in the late twentieth century was remarkably matter-of-fact, even for a place as far off the beaten track as the great red island—Madagascar.

These days, no country on earth is more than a couple of days' trek from anywhere, and for my journey to Madagascar I got not a single inoculation, bought no special medicines, took no particular precautions, but simply boarded the plane as I would have for a trip to Ohio. The flight across the Atlantic to Paris was typical of any night flight: sleep, eat, pee, stretch, disembark. A boring five-hour layover in Charles de Gaulle International Airport was followed by a spectacular flight from Paris to Antananarivo. The view quickly became scenic as we tracked southwest out of Europe. First it was the rugged, expansive, snow-draped French Alps; then down the western side of the spine of Italy; and finally a peek at Sicily and the Mediterranean. When we crossed the coast of northern Libya near the Gulf of Sidra, the sun was beginning its descent to the horizon. I began to see the striking colors and patterns and geometries of these northern African desert lands. The land itself, with no visible evidence of humankind, was a tapestry of hues below, the rich wavelengths cast by the low sun elongating the shadows and sanguine tones.

The nearshore interior was brick red, unvegetated, but punctuated by tiny white puffy clouds low over the land. Next came a paler buff landscape that before long graded into huge complexes of sand dunes whose sharp ridges created a geometric pattern across the devegetated landscape. It is remarkable to think that these treeless desert lands were, half a million

Preceding page: A small chameleon in Madagascar

years ago, humid tropical forest lands, with now-extinct primates and a rich diversity of plants and animals—a far cry from the impoverished biota that populates the interior of northwestern Africa today.

If the reader is wondering what happened to the rainforest, the un-surprising answer is . . . global climate change. It is not a new phenomenon: climate change is the rule, not the exception. And climate change was the rule long before humankind came to dominate our earth or to infuse our atmosphere with greenhouse gases. Climate change, extinction, and specia-tion have been acting in concert for many millennia. Past changes in the climate of northern Africa certainly caused local extinction pulses. These have been well documented by paleontologist Scott Wing, who has written of the Koobi Fora flora and fauna—a now-vanished humid tropical world in northern Africa.

We passed over clouds, then over a vast open dry savanna broken by the green floodplain created by a large river course (the upper Nile?). To the west were immense dark pillars of storm clouds that flashed with lightning. Darkness fell, and the full moon began to rise over the interior East African hinterland. Unexpectedly heavy El Niño rains had been falling over the rift lakes region, creating floods in the back country that so often suffers from drought. That night I could look down on the East African landscape and see the cold white shimmer of moonlight reflecting off one of these lakes as well as adjacent flooded savannas. As we approached the coast at Dar es Salaam (Tanzania), saturated wetlands gleamed in the moonlight. We could glimpse the lights of Zanzibar off the coast, which cheered my late-night reverie. Ahead lay the dark bulk of Madagascar across the Mozambique Channel, where deep below lurked the antidiluvian coelacanth fish.

We touched down before dawn on the tarmac at Antananarivo, the upland interior capital of Madagascar. At predawn it was pleasant and relatively quiet at the airport, which was small and quaint. Navigating through baggage and customs and immigration presented no insurmountable problems to a first-time visitor with no grasp of the dominant languages (French, Malagasy).

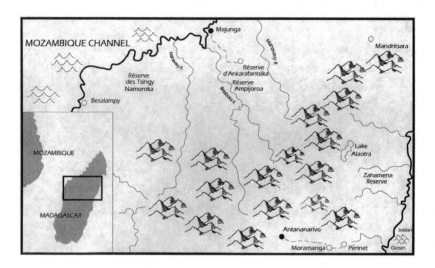

The one noticeable difference here was that the arriving passengers broke
into waiting lines with more bravado than in the United States. I found myself
muddling through, mumbling, nodding, hoping that my two bags would roll
out on the baggage cart undamaged.

It is remarkable how often this process goes without a hitch. In thirty
years of international travel I have never lost a bag, although in many cases
bags have been temporarily misrouted, causing some inconvenience but no
real hardship. (Once, in Delhi, I did have to spend time in the back room of
a customs office because the customs people had failed to see my valid visa
right there in my passport. They almost convinced me I had no visa!) All in
all, travel these days is wearying but not terribly difficult. The professional
"travel-adventure" writers are often forced to fall back on that old standby,
tall tales, to make their generally humdrum story of traveling into the "far
wilds" more interesting to the reader.

In 2008, the main issue, of course, is the absurd and inefficient anti-
terrorist structure, driven by the keen paranoia of the U.S. government bu-
reaucracies with jurisdiction over air flight. This mania seems to drive the
caution that plagues all the Western nations. Away from the beaten track,
the restrictions drop away and one can recall the situation only a few years

back, when life was "normal." Certainly flying in 1998 was a lot more pleasant than it is in 2008.

❧❧

I immediately liked "Tana"—Antananarivo. Though a big, sprawling city arrayed over a series of hills, it has a small-town feel because of the narrow winding streets and abundant greenery. It is on a high plateau (about 4,000 feet) and thus is cool and moist for much of the year. Rice grows in paddies in and around the city, giving it a pastoral feel. And most of the buildings, with a few hotels the egregious exceptions (the Hilton, the Palace), tend to be small and low and inviting.

I attended the big USAID meeting and on the following days made official visits to a wide array of envirobureaucrats. It was the necessary routine: lots of phone calls to make appointments, then lots of frenetic taxi rides to the various offices and hotels. Everyone was polite and friendly and at least feigned interest in the *Forest Gardens* program I was pushing. At the time, in 1998, Madagascar was one of the darlings of international nature conservation, so everyone was here, doing projects. These projects were conducted by nongovernmental institutions like Counterpart or Conservation International or Chemonics or CARE. The project implementers were here because this was where the money was—the international aid money, provided by the likes of USAID or UNDP (United Nations Development Program) or World Bank (International Bank of Reconstruction and Development), or GEF (the Global Environment Facility). What a dance! And one that the funders most definitely led.

Here is how it worked. The international donor agencies visited the local government agencies and high-level bureaucrats in Madagascar and made offers of foreign aid. The Malagasy bureaucrats made demands on the aid agencies, and a back-and-forth negotiation ensued over what the form of the aid would take, and how the conservation would be achieved. The result was complex agreements between government and donor agencies, which in turn led to a series of competitive funding initiatives. The conservation and consulting groups sprang into action, with a mad rush to form multi-institutional alliances that could deliver the goods—conservation and rural development on the ground in exchange for grant funds. I was in the midst of the alliance-building phase, looking to sell Counterpart and its *Forest Gardens* system as a key component in one of the alliance teams who would make a bid to USAID and perhaps to other donors. That was my business mission in Madagascar. My other mission was self-education. I needed to understand Madagascar and its environment in order to be an effective salesman. To learn, I needed to see it firsthand.

With the weekend approaching, I planned a quick visit to the nearest nature reserve, popularly known as Périnet (more properly Parc National d'Andasibe-Mantadia). A drive of a couple of hours brought me to the humid highlands that dominate the landscape of the eastern side of Madagascar. The asphalt road took us through lush green hill country, most of it entirely lacking in original forest. In its stead we saw an abundance of paddy rice, rich and glowing green under the bright upland sun, the color defying reality in its intensity. Interspersed between the rice paddies were stands of eucalyptus—pure monocultures of an Australian tree that is harvested a few short years after planting to be converted to charcoal. It provides the fuel that powers the home fires of rural and urban Madagascar.

Charcoal is half the story of Madagascar's devegetation. The central upland plateau has been devastated by the population's demand for fields for rice and for the necessary charcoal to cook it. The only remaining forests and woodlands nestle in isolated patches around the island's edges. Along the wet eastern mountain chain the forests are more abundant, and along the western (dryer) side woodlands hang on in isolated river valleys. Here Madagascar's remaining wildlife hides from the ever-growing populace.

The only advantage of the current boom in growing eucalyptus for charcoal is that it is a "sustainable" process. The trees are grown on open ground, so no forest is sacrificed. Still, it is a bit like growing corn: the process does little to support nature or biodiversity. Elsewhere in the world it is common to see original forests cleared to make way for monocultures of exotic tree species (various species of eucalyptus, *Pinus caribea,* Big-leaf Mahogany, and so on). That is deplorable. For in removing all the natural vegetation to make way for the plantation, entire natural systems comprising tens of thousands of native species are wiped out, replaced with unnatural systems made up of one or two economically valuable species and a handful of weedy tramp species that can make it in this human-managed habitat.

At Périnet, I was put up in a jungle lodge named Hôtel Feon'nyala at the edge of the reserve. The lodge was small and rustic, a perfect antidote to crowds of Antananarivo. While I got settled in, Parson, my driver, picked up Narina, my assigned local nature guide. We went out for an afternoon look-around, and within minutes, we were listening to the wonderful territorial howls of the Indri, Madagascar's largest living lemur species, one of this island's specialty primates.

Monkeys and great apes inhabit Africa. Lemurs, by contrast, enliven Madagascar. For me this trip would be special because of the lemurs, plus the vangas (the avian version of lemurs here), as well as the comical chameleons. This afternoon's amble was a low-key, pleasant introduction to the natural attractions of Madagascar, giving me a taste of what was to come. Narina showed me several endemic bird species, including the demure Red-tailed Vanga, which looked somewhat like a jazzed-up House Sparrow. It was my first encounter with this special bird family that only inhabits the red island.

The evening at the little Feon'nyala lodge was paradisical. My peaceful evening meal was followed by a quiet sit on the wooden veranda of my little A-frame bungalow, with a serenade of katydids, frogs, a scops owl—Sirius glowing, and the Big Dipper draped across a black sky that shimmered with stars above the mysterious forest.

At dawn I could hear the bleats of the Indri, calling from a distant ridge in the adjacent forest. From the stool on my front porch I watched unidentified birds moving through the vegetation along the forest edge that glistened with raindrops. Parson came to pick me up and we waited as Narina strolled down the road to us with a cheerful grin and a spring in his step.

Black Bulbuls, a species I knew well from northeastern India, were

The Indri—the largest living lemur species and
the flagship of Périnet Reserve (photo courtesy of
Russell A. Mittermeier)

omnipresent and noisy in the canopy vegetation, now lit with sunlight. It
is remarkable how many species from India are also found in Madagascar,
a reminder of the two regions' strange biogeographic linkages. Madagascar
and India once were contiguous with no Indian Ocean to separate them, but
that was a long time ago.

That morning the first thing we three did was venture out in search of
the Indri troop. We found a family of three in the forest, by a roadside at the
park boundary. An adult pair and a two-year-old youngster were stretching
tentatively, giving us groggy looks as they prepared for the day ahead. They
perched quietly in the tree, then started to move off into the forest interior.
Not many monkey species can come close to lemurs in terms of splendid
furriness, soft beauty, and simpatico. Lemurs seem like small people out of
time, from another place and another world—which of course is exactly
what they are.

Many of the lemur lineages are reminiscent of other mammal forms
around the world. The Indri looks a bit like a slimmed-down panda. By

contrast, the prototypical member of the family, the Ring-tailed Lemur, looks a lot like a Coatimundi of the Neotropics. A Bamboo Lemur resembles an owl monkey from South America. The tiny mouse lemurs are like cute versions of the tarsiers of Southeast Asia. The Hairy-eared Dwarf Lemur looks like another neotropical primate, the Pygmy Marmoset. The Fat-tailed Dwarf Lemur looks a lot like a Slow Loris of Southeast Asia. The fork-marked lemurs are spitting images of the Australasian trioks (marsupial squirrels). The Black-and-white Ruffed Lemur is a gracile version of a Black-and-white Colobus monkey from central Africa. Finally, the long-limbed and slim sifakas closely approach the spider monkeys of Central and South America.

Like so many special lineages around the world, today's seventy lemur species are but an imperfect remnant of a rich historical radiation. We now know that an array of extinct species, including the slothlike *Babakotia* and the giant *Megaladapsis,* were large headed and may have approximated koalas in form. The now-extinct *Hadropithecus* probably looked a lot like a baboon. And, the giant *Archaeoindris* may have reached 400 pounds and perhaps resembled a giant ground sloth.

Eight genera and fifteen extinct species have been described to date. That is a shocking amount of lost biodiversity, especially considering that these are recent extinctions, having taken place in the two thousand years since humans arrived on the island. These were mainly larger forms that were vulnerable to human predation for the stewpot. Other creatures followed them into oblivion: the huge, ostrichlike *Aepyornis* or Elephant Bird (probably the largest bird that ever lived, weighing 600 pounds), a pygmy hippopotamus, and giant tortoises (reminiscent of those of the Galápagos).

While regretting the loss of the now-extinct forms, we can still celebrate today's incredibly diverse lemurs. To the first-time visitor to Madagascar, the lemurs are a mammalian feast for the eyes. Eleven species of lemurs live in and around Périnet. Comparable numbers can be found elsewhere on the island. Each natural habitat is home to an array of these docile primates, found nowhere else on earth. By comparison, most of the monkey-rich habitats support no more than five or ten different monkeys.

So Madagascar is remarkably rich—extraordinary, really. Here at Périnet

I encountered six species. Aside from the panda-like Indri (black-and-white, tailless, with those tawny staring eyes and big ears), I saw the lovely and gentle Eastern Lesser Bamboo Lemur (fuzzy gray-brown and demure, with a fluffy brown tail) and the Brown Lemur (a bit like the cuscus I would expect in New Guinea, but more endearing and intelligent looking). How little we know about Madagascar's lemurs is evidenced by the fact that the first edition of *Lemurs of Madagascar* in 1994 recognized thirty-two living species. The revision of that guide in 2006 featured seventy-one species!

We left the Indris to chase some birds. Nuthatch Vangas played about in the canopy, the male looking for all the world like a sky-blue nuthatch. During the morning I encountered four additional vangas: Hook-billed, Red-tailed, Tylas, and Chaubert's. Vangas are Madagascar's answer to bush-shrikes as well as a dozen other Old World families. They are strikingly patterned with two or three blocks of color—white, black, blue, or brown. They are easy to identify and fun to watch. Some have weirdly shaped bills. Others closely approximate the forms of birds well known in other parts of the world (for instance, Velvet-fronted Nuthatch in India compared to Nuthatch Vanga in Madagascar). Seeing unique animal and plant lineages is what nature tourism is all about, is it not? I was definitely playing nature tourist here in Périnet.

The splendid diversification of the lemurs and vangas indicate that Madagascar is a land apart. Originally a piece of the vast southern continent Gondwana, Madagascar broke from the rest of Africa about 160 million years ago and separated from India about 90 million years ago. It was the resulting isolation that allowed the evolution of these endemic lineages.

The forest at Périnet is small and low and disturbed, but green and lovely, reminding me of Papua New Guinea's Varirata National Park in the wet season. Remarkably, the sounds and smells of the forest are all quite reminiscent of Papua New Guinea. Tropical humid forest has a character that makes it special and recognizable. What set Périnet apart from New Guinea was that in our brief rambles we found a Common Tenrec, Madagascar's version of a hedgehog, a big male chameleon, and lots of birds. New Guinea is much less generous in sharing its animal life.

Maurice, the bird specialist and guide, took us out on our second morning in Périnet. We tramped up and down old French-period logging tracks (the French were dedicated timber extractors) through mossy forest in search of Madagascar's avian rarities. Maurice played a tape on his decrepit tape recorder to lure in a pair of Madagascar Crested Ibis. The birds slowly made their way through the forest right up onto the track, giving their wretched vocalization, *gro gro gro gro gro*. We found a Forest Rail, a Short-legged Ground-Roller, a Velvety Asity, a White-throated Oxlabes, and we even heard the elusive Madagascar Serpent-Eagle. Madagascar has lineage after lineage of mysterious endemic birds, which for maximum appreciation must be learned prior to arrival here. It was hard to get used to these entirely new species that can be found nowhere else on earth. In the afternoon, Maurice lured a Flufftail out onto the paved road—such a tiny and exquisite little rail! We bumped into a troop of Brown Lemurs by the roadside at the end of the day. They were playful and active, moving about through the open canopy. It was fun to watch an adorable youngster, barely independent of its parents, at play. Babies at play, no matter the species, are invariably endearing.

What became clear to me was that this small reserve at Périnet was an island of nature surrounded by human-managed eucalyptus or pine monocultures, rice fields, and rural habitat of small huts, carrot and leek gardens, and plantations. There was not much left for the wildlife. People lived everywhere and managed all of the available land for subsistence or their meager enterprises. This heavily modified agricultural hill country reminded me so much of the interior of southern Sri Lanka—Bandarawella in particular. The combination of eucalyptus and exotic pine plus a range of market gardening was remarkably similar in the two locations. Both habitats badly needed a big infusion of permaculture cropping to create diversity rather than homogeneity. The *Forest Gardens* system was working well in southern Sri Lanka, and it certainly would be a valuable addition to the Madagascar upland landscape.

On my last morning in Périnet, Maurice took me out of the reserve to visit the Maru Mizaha State Forest. This government-owned block of managed forest had a very different feel from Périnet. We took a long walk up a

ridge following a former logging track, much like the old Bulldog Road south
of Wau, in Papua New Guinea. I was struck by the similarities of vegetation—
the bracken, the tangles of *Dicranopteris* fern, the scrambling bamboo,
and even a weevil with a weird growth of algae on its back like one finds
along the Bulldog Road. The southern humid tropics have a "jizz" that is
recognizable.

As a state forest, Maru Mizaha had more of the feel of the "real world"
of the Madagascar countryside, so was of interest with respect to its potential
for Forest Gardens agriculture. I could hear the woodsman's ax in the distance. I looked down the slope to a small clearing of tavy (dry land) rice, and I
could hear troops of Indri calling to each other from different forest patches.
At the high point of the track, we sat for an hour at an overlook created by a
recent landslide. We scanned the forest for sifakas. We watched Greater Vasa
Parrots float up and down the valley. We marveled at an iridescent Urania
moth that fluttered across the opening.

The integration of nature and rural life that I witnessed on that ridge
was not at all discordant to me. The gentle mix of natural and rural is something that seems to soothe the human senses, whether in the rural United
States or in rural Madagascar. I found the mix of small-scale agriculture,
hamlet, mountain, and forest particularly comforting. There must be some
ancient gene that generates a human search image for an ideal environment,
and this was it. By comparison, a vista entirely devoid of the human touch
is majestic but often cold and even a bit threatening (the "forest fastness").
What is difficult, in Madagascar as well as rural Maryland, is how to balance this mix of human and natural. Demography tends to push human
landscapes to the "all-developed" option. Does anyone really like looking
out at a hillside overgrown with a monoculture of row-planted vinyl-paneled
tract houses? Yet it is the ever-more-dominant habitat in my homeland of
suburban central Maryland.

❧❧

On Monday afternoon Parsons and I drove back to Antananarivo through the
hilly open countryside. The sky was deep blue and the countryside was ever

Terraced paddy rice and local mud-daub homes in the highlands of Madagascar.
The rice fields glowed in the highland sun.

so green. The paddy fields were an unworldly verdant hue that contrasted
perfectly with the ocher and whitewashed stucco houses that bordered the
fields. The patterns were reminiscent of a painting by Cézanne. His favored
habitats for painting were overdeveloped, overcleared, human-dominated
French rural environments, but they were still remarkably beautiful, very
pleasing to the eye. In the same manner, this rural upland landscape provided
ideal camera images for exposing on Fuji Velvia film, whose specific mix of
surface chemicals reacted perfectly to the subtle and delicate pastels of the
tropical countryside.

   After several more days in the capital city, I headed off to visit Madagascar's northwest. Looking down from the jet en route from Antananarivo
to Majunga was a shock—nothing but bare red earth and a myriad of deep
erosional gullies. It is on this aerial transect from central to northwestern
Madagascar that I witnessed the full measure of environmental destruction
that had been wrought. The Malay-stock migrants who settled in the interior uplands of Madagascar and formed a series of kingdoms decimated the

natural environment of these once-fertile uplands. The streams and rivers that drain these burned-out lands still often bleed a deep brick-red color, different hues mixing as they coalesce toward the coast. It is a ghastly but breathtaking sight as the great watercourses pour their terrible bloody-hued sediment load into the Mozambique Channel, swirling in hideously beautiful blooms into the blackish ocean waters between this island and mainland Africa.

The last fifteen minutes of my flight to Majunga began an interesting environmental transect of habitats. Leaving the interior upland wastelands, we moved over thick, dry, deciduous forest and riverine gallery forest, from there to mangrove (the deep green contrasting with the brick red of the rivers), and finally to the coastline and Majunga town, with lovely beaches and a diversity of periurban settings. As we landed I could see giant Baobab trees along the town's broad streets.

Having departed Tana in the misty cool rain, I arrived at the coastal port of Majunga (Mahajanga) in terrific sun and heat—traveling from upland humid climes to coastal semidesert. For here, in the seasonally dry zone, we found oppressive heat that baked any land stripped of its native trees. Here in town, a grove of mangoes was the only place where we might have stopped for shade and a bit of respite from the blazing noonday sun. It happened that my host here also served as director of the Majunga airport. He and his team efficiently organized all aspects of my next field trip into the interior. They got me food, drinks, a car, a driver, and a guide. In no time we were on the road to the Ankarafantsika National Park, where Conservation International had been running an ecotourism project and assisting with park management.

Our drive took us through hilly country with lots of interesting sedimentary outcrops. Having read of the paleontological history of this part of northwestern Madagascar, I kept thinking how exciting it would be to stop and do some digging for fossils. Various Mesozoic dinosaurs had been described from this area in recent times (*Majungasaurus,* for instance). After nearly three hours of driving, we dropped into the low interior basin that constituted the 330,000-acre park. It was very green now, at the beginning of the local rainy season in this interior location.

At the park I was met by my local hostess, Madame Mami, who gave me a place to camp under a thick canopy of planted trees (protected from the noonday sun). I set up my tent on flat, bare ground, which made me worry about flooding should heavy rains come at night. Also I learned about a remarkable tree here that actively shed hundreds of large sharp thorns, which were scattered all around on the soft ground. They gathered like tacks on the bottom of my sandals. But the thorns were quickly forgotten when a troop of Coquerell's Sifakas appeared in the trees at the edge of the campground, leaping playfully from trunk to trunk. These graceful brown-and-white primates, with their large eyes and intelligent faces and lithe bodies, allowed me to approach them to within about 40 feet—confident, but not tame.

After a quick lunch my guide led me off into the forest, which was unlike anything I have ever seen in the tropics. The whole area was dominated by low sandhills; the soil was pure sand. The deciduous forest, green and lush at this season, was gnarly. A bit like North Florida without the pines or chiggers . . . It was, however, rich with wildlife. We bumped into troops of *Avahi, Lepilemur,* and *Propithecus*—lemurs by the dozen, and all at close range. We encountered six troops of the Sifaka, some of which allowed me to get within a couple of yards of them. Madame Mami was an excellent naturalist, showing me a big ground cuckoo (Crested Coua), a cute green gecko, a large and picturesque chameleon, and a remarkable recumbent plant that literally lies flat on the ground, standing no more than an inch high. The only thing that Madame Mami lacked was English. My French was poor, so we spent a lot of time waving our hands and searching for the right word in each other's language.

In the late afternoon, the clouds rose high into the sky and a thunderstorm broke over the campsite. Rivulets rushed this way and that, but my tent stayed dry on the inside. A cold bottle of beer in combination with some salted cashew nuts provided a satisfying cap to a day of travel and discovery. My driver cooked a simple, tasty dinner, and I was asleep soon after dark.

After a long night in the tent (dark comes early in the tropics) I was up at five-thirty for a quick breakfast followed by a trek into the stunted forest. The birds were quiet. Madame Mami gave me to understand that all the birds were nesting and quiet at the moment, thus difficult to see. So instead we played with tenrecs. These primitive insectivores are weird and marvelous. Much like tiny hedgehogs, tenrecs are tailless, coarse-haired, pointy-snouted beasts that forage for invertebrates on the floor of the forest. We found a number of the buffy-colored adult tenrecs this morning, each with a crèche of adorable stripe-backed babies. The tiny young huddled together and moved about as a single mass, with the parent standing guard. Upon our approach the young stopped and the parent defended them. It was thus possible to see them up close and appreciate their totally wonderful qualities.

Otherwise the morning was pleasant but quiet, enlivened by encounters with some fascinating birds: Van Dam's Vanga, White-throated Mesite, Red-crowned Coua, Coquerelle's Coua, and Sakalava Weaver. We found reptiles too, including a multicolored skink and a large snake.

At the end of the day I took a quiet supper of noodles, bread, beer, and pineapple. Without anywhere to hang out except for my little tent, after dark I wrote in my diary, dressed only in a bathing suit. The rain pelted down, and the tent was hot and humid. Sweat coursed down my back and forehead. The tent was a sauna, which had retained much of the noonday heat. That temperature was boosted by the heat my body was generating from the digestion of my supper. It was sweet misery, though I knew that in an hour I would be cool and comfy, nestled under my sleeping sheet.

The next afternoon I visited the nearby lake to look for the Madagascar Fish Eagle, a local endemic that looks a bit like our Bald Eagle, so familiar from the Potomac environs around my home. The eagle was quickly forgotten, though, when I bumped into the supersighting of the trip—a sociable flock of some thirty Sickle-billed Vangas. These white-headed creatures, with their long pale-blue decurved bills, were reminiscent of the Buff-tailed Sickle-bill I had studied in Papua New Guinea for my doctoral research. The latter species, however, was antisocial and highly territorial. The vangas moved slowly and noisily through the canopy of the edge of the forest. I stayed with the flock for more than forty minutes, wondering what advantage this strong sociality afforded the birds. That evening I was visited in the campground by a human-habituated Brown Lemur, an entrancingly fluffy creature that allowed me to cuddle it. It was friendly but did not make a nuisance of itself. What gentle creatures lemurs are!

The next day I had a variety of discussions with local staff about the park and the project run by Conservation International. The talks highlighted just how difficult it was to develop a workable conservation project under the local conditions. The villagers complained about the project mainly because establishment of the park had cut off access to many natural resources on which the people had long depended. The local population was not allowed to hunt in the forest, fish in the adjacent lake, or collect *Manihot* tubers in the forest. The local communities were poor and they found these strictures cruel. The villagers produced limes and mangoes as market products. Because of seasonal overproduction little cash income resulted for all their effort. It was clear that considerable diversification of their agricultural production was needed to achieve a more stable and productive rural economy.

The opportunity for nature tourism here was strong, but there were no facilities—no lodge, no restaurant, no lavatory. Poverty was so widespread, opportunity for improvement so terribly limited. Aside from small-scale agriculture, which is predominantly subsistence cropping, there was little else. The employment of eighty local people as staff of the park project was probably the major economic news, a real indication of how little opportunity existed in the countryside. Housing in the local villages was appalling; mud-daubed houses or thatch huts, tiny and miserable, baked in the hot sun. Many villages had little sheltering vegetation; the ground was merely pebbles, the topsoil having been lost to the powerful erosion.

Even the most impossible hillsides showed evidence of government tree-planting. But ensuring the success of such reforestation efforts is an arduous battle, with the lack of soil, the long dry season, and the impact of dry-season burning. Villagers plant fruit trees, but for some reason these little plantations are not located near the houses, leaving them in full sun. It is disturbing indeed to see the villagers arrayed along the roadsides, selling bottled mango and lime in recycled water bottles.

The next day I was up early and out along the road to see if I could find the Sickle-billed Vanga flock. It was right where I had left it the night before. The birds were noisily waking and beginning their day. Dense fog rose off the lake, and the vangas looked rather like spirits in the trees. Madame Mami took me for a final traipse through the forest, and we located more of the little spiny tenrecs. We discovered that if you squeak at the tenrec family the whole group will, on command, march right up to you. I photographed a Sportive Lemur in its roost in a tree hole, and saw a troop of Brown Lemurs and a single Sifaka. Last of all, Mami pointed out a Cuckoo-Roller making its display flight in the cerulean sky high above the forest. I now had seen all of the bird families endemic to Madagascar.

At eleven in the morning, we began the drive back to Majunga. I had to spend the night there before the next morning's flight back to Tana, the capital. The little coastal outstation of Majunga was bereft of infrastructure, the prototypical sleepy tropical provincial port. Everything was old and crumbling—shabby chic! The ancient buildings and old streets were long past their prime. The town barely remained except for the people and the cars. Wide streets led nowhere. Dusty, sleepy, decaying, endearing Majunga. My lodging, the Hotel de France, was the best in town. It was survivable, but certainly would not make a Michelin guide. At least the staff ran my ticket down to the Air Madagascar office for reconfirmation, so I could not complain (I avoided a trip in the killing noonday sun).

After a couple of days back in Tana, occupied with official meetings, I headed out to Nairobi and then Sri Lanka, in my role as envirocarpetbagger. The most memorable moment in the Malagasy capital was watching CNN in horror as the Clinton-Lewinsky scandal erupted across the world's media. At that moment, America was serving as a global sitcom.

My international flight home routed me right over Ankarafantsika Park, which from high above was a lush green refuge surrounded by the red scars of devegetation. We touched down briefly at Moroni, capital city of the Comores on the new volcanic island of Grande Comore. What a contrast with ancient and continental Madagascar, for here was an island being born before our eyes. Great volcanic cones rose to 7,000 feet, and young black lava flows were everywhere. Volcanic cliffs marked the coastline. Raw fresh-

minted volcanic land thrust up from the tropical sea, for the Comores mark
an active tectonic plate boundary, exemplifying the majestic earth processes
of destruction and regeneration.

Today, the international development agencies and conservation organizations are hotly debating poverty alleviation versus environmental protection.
The World Bank and the other multilateral aid institutions are focused on
poverty alleviation, and the level of investment in the environment is lagging.
What exactly is the relationship between poverty and the environment?

Let's examine the poor inhabiting the lands around Ankarafantsika
National Park. The natural lands and forests colonized by the local people
have provided these communities with many economic benefits. However,
the local populace has, over time, degraded these lands, clearing them for
agriculture and then overworking them, producing soil erosion, local pol-
lution of the streams, devegetation of critical catchment areas. The problem
in northwestern Madagascar is exacerbated by the long dry season and seasonal burning. In the tropical dry zone, old-growth dry forest is rich but
exceedingly fragile and easily destroyed by humans. There is little hope for
regeneration, mainly because of the impact of chronic fire.

What the world forgets is that human communities are subject to the
same environmental constraints as other organisms living in the environment.
It is a matter of carrying capacity. Small populations of humans in marginal
lands tend to subsist productively; but once populations cross a threshold,
the ability of the environment to support the population breaks down. Most
rural poor are poor today for this very reason: too many individuals per
square mile of habitat, with ongoing degradation of the resource base.

Poverty is an environmental issue, first and foremost. We tend to ignore
this fact, mainly because we only focus on the poor once they are trapped in
what comes to be a "human environment" rather than a "natural environment." For instance, in parts of the Sahel concentrations of poor have settled
onto land that is, in essence, entirely depleted of the essential resources
needed to support a healthy community: clean running water, rich topsoil,

sufficient annual rainfall, and adequate forest cover to protect critical catchments. These extreme situations are usually caused by a breakdown of civil society and the mass movement of people plus the failure of land tenure systems.

The counterexample is the rural situation in Papua New Guinea, where land tenure is strict, population is low, and the natural resource base is strong. In this situation, the rural communities are not poor, and the environment is providing the services that allow them to thrive.

How do protected areas fit into the equation that includes protection of critical environmental services, biodiversity, and the rural poor? It is another contentious issue, especially where rural populations are high and where local populations lobby to have access to resources within protected areas (as with Ankarafantsika). From the viewpoint of a conservationist, the development of national networks of strictly protected areas (parks and reserves) is absolutely necessary for the long-term well-being of each nation. It needs to happen before expanding rural populations overrun and degrade every last tract of natural habitat. The healthiest nations are patchworks of populated land and protected habitat.

The example of city parks is instructive. Creating Central Park in New York City would be utterly impossible today. The land in Central Park could be sold to developers for many billions of dollars. Yet even with a burgeoning city population, the value of this great park is not questioned. No one would dare recommend that the park be converted to high-rise apartments, even though New York has a chronic housing shortage. Everyone recognizes that Central Park makes Manhattan a better city, even though it took nearly a thousand acres off the tax rolls and reduced the amount of buildable land on the island. The remarkable and intense city lifestyle is made bearable because of that critical green space in the heart of the city.

For some reason, when a park is put in a rural area, the nearby residents show less appreciation for it, but the same principles apply. Having a protected area in the vicinity has been shown, over and over, to improve the local economy and the quality of life. Watersheds are protected. Tourists are attracted. There are even spillover effects with huntable wild game.

The resentment of rural poor living near a protected area usually can be quickly defused by two simple actions. First, make certain the local community has a voice in an appropriate stakeholders' forum overseeing management of the park. Second, be flexible about sustainable use of park resources (managed through debate within the stakeholders' forum). Local stakeholders need to participate in park management, and park regulations need to be flexible and fair.

Sitting in the plane on the tarmac of the Moroni airport, I thought back to Madagascar and what I had seen in my brief tour. It is a land that provides great value for the well-heeled visitor in search of remarkable wildlife and scenery, yet it is so poor and so needy. This is one of the paradoxes of tropical ecotourism. I was on official business, and yet I was, to all intents, an ecotourist. Did Madagascar benefit from the francs I spent in-country? Did some gain trickle down to the rural village poor? I made my initial travel and hotel reservations with a prosperous specialty travel agent in San Diego. I made additional arrangements in Tana. I don't believe that much of this spending made its way to the communities around the two parks I visited, though more stayed in Périnet, certainly, because of the clever local association of naturalist-guides there.

One might suppose that local residents should directly participate in providing tourism services to visiting foreigners and thereby earn tourism dollars. But there are barriers to this option. The local residents possess neither the capital nor the training nor the know-how to provide the nature-tourism opportunity desirable to the tourists. It is much better organized and arranged by outside interests, by people who are culturally more like the tourists than the local residents and who understand what the tourists want. So a culture war arises. The local residents resent the outside tourists, who are the cause of the demand for effective nature conservation, which restricts the local opportunity for resource exploitation.

As wild sites become more and more popular, and as nature-tourism opportunities expand, the heavy traffic of visitors destroys the precise qualities that made the resource attractive to those tourists. Thus the traffic jams on the small roads in the Adirondacks reduce the quality of the wild experience. The push to build larger roads reduces the traffic jams but leads to the very types of development that the tourists are trying to escape. In other words, wild-nature tourism can bear only minimal tourist traffic before it begins to degrade the exact resource sought by the tourists. If only low traffic and few tourists can visit, the economic incentive to conserve the wilderness resource is diminished. Extractive development is probably a more economically productive activity, with greater local benefits for the rural communities, if only for the short term (until the forest is logged over or the mine played out).

The key to success is to foster locally owned, locally operated tourism that is managed for the long term, and that provides sustained benefits to local stakeholders, not just to outsiders. In Madagascar, time will tell whether economic opportunities can develop in a place like Ankarafantsika in ways that protect the wild resource while providing a growing return for the rural poor. The naturalists' cooperative in Périnet is an example of what can happen with some local initiative. Supporting such useful steps forward should be the mandate of governments and of international conservation organizations.

CHAPTER TEN

The Lost World

It seemed so incongruous that birds of paradise,

so brilliant and delicate,

should live in such moist, giantesque,

subdued surroundings.

E. Thomas Gilliard,

"To the Land of the Head-hunters,"

National Geographic (October 1955)

HAVE YOU EVER WONDERED WHERE the most remote place on earth is? It is the Foja Mountain range in western New Guinea. As mentioned in the Introduction, my 2005 field trip to this isolated range was a revelation—no villages, no litter, no logged-over or gardened areas, no roads, no walking tracks, no hunting, not even any sounds of civilization except for a once-a-week passenger jet flying high overhead.

The Foja range was the place that helped me to better understand the links between plate tectonics, mountain building, and the evolution of new species. Finding the Foja Mountains to be a place rich in undescribed species emphasized how little we know of some corners of Planet Earth. And ironically, visiting the Fojas taught me a stark lesson in the relentless process of extinction that stalks even species hiding in the vast rainforest lands of New Guinea.

❦

Isolated from other mountain ranges by the Mamberamo Basin, and rising to an elevation of 7,250 feet (higher than Mount Washington), the Foja Mountains represented the last unsurveyed major mountain range on the island of New Guinea. By the late 1970s New Guinea was, in fact, pretty well studied biologically, but some corners were still unvisited. The Foja range was at the top of this short list.

❦

The Fojas were special because of three scientific mysteries: a bowerbird, a bird of paradise, and a tree-kangaroo. The two bird mysteries took us back to the Victorian era, the time of aristocratic "cabinet naturalists." Adventurous souls braved the jungles to collect new plants and animals to be described and named. In Victorian times, naming (especially of butterflies, beetles, birds, and orchids) was a competitive and popular "sport" among the aristocratic elite of Europe. The cabinet naturalist who nabbed the first specimen could publish a paper describing the new species, giving it a name, and immortalizing himself at the same time.

Here's how the system worked. The naturalist would receive a wooden crate from, let's say, the Far East. Cracking open the box, he would find the

Preceding page: The Golden-mantled Tree-Kangaroo

crate filled with dried bird study skins, each bearing a collector's label. The naturalist would shuffle through them, looking for something that might be "new," something that had never been described by Western scientists and had not appeared in any paper or checklist or monograph.

Finding a set of specimens of an apparently new species, the naturalist would happily set about writing a paper describing this novelty, perhaps naming it after the Queen. Thus did the naturalist Fraser in 1844 name a huge blue ground-dwelling pigeon from New Guinea *Lophyrus Victoria* in the *Proceedings of the Zoological Society of London.* Fraser thereby immortalized Queen Victoria and at the same time associated his own name forever with that wonderful ground-dwelling pigeon (because a scientific name is associated with the namer in the technical literature even today; hence we have today's emended name *Goura victoria* (Fraser)—the Victoria Crowned Pigeon. In the science of taxonomy, the creature gets a permanent name and the namer gets the credit for the naming. Each species in nature has *one* accepted species name that becomes the permanent epithet for that species.

Thus, in 1895 Lord Walter Rothschild, on his grand estate of Tring, west of London, opened a box from the Dutch trader Duivenbode that, among other things, contained three specimens of a strikingly beautiful and novel gardener bowerbird of the genus *Amblyornis.* Although darker and browner than the other known species of the genus, it most remarkably sported a gorgeous golden-yellow mane that stretched from its bill to the middle of its back—a long silky erectile crest that makes this male one of the most beautiful of the bowerbirds.

Rothschild wrote up the description for publication in his own scientific journal, *Novitates Zoologicae.* He named the bird *Amblyornis flavifrons* ("dull bird with a yellow front"). For the bird's range, Rothschild noted "Dutch New Guinea." Herein lay the mystery: the specimens had apparently been collected by indigenous hunters from an unknown location. Given the size and inaccessibility and undeveloped nature of western New Guinea, it was no simple task to locate, after the fact, the precise home of this lovely bowerbird.

Similarly, in 1897 the German ornithologist Otto Kleinschmidt described a new six-wired bird of paradise, *Parotia berlepschi,* from trade skins in the private museum of Hans von Berlepsch (probably also from the trader Duivenbode). This one was a velvety-black bird with six erectile headwires (in lieu of a crest), each having a little spatulate tip, an iridescent throat patch of metallic feathers, and striking white erectile flank plumes. As with the bowerbird, the *Parotia* apparently originated from some unknown location in western New Guinea.

From here the story is told by E. Thomas Gilliard, in his book *The Birds of Paradise and Bower Birds.* Gilliard noted that at least a dozen expeditions were mounted to New Guinea in the subsequent sixty years in search of the homeland of the bowerbird and the bird of paradise. These expeditions, led by the likes of Bruijn, Mayr, Gilliard, Bergman, and Ripley, all failed to find the two lost avian treasures. They scoured a number of poorly surveyed mountain ranges, all to no avail.

Following in Gilliard's footsteps, UCLA's Jared Diamond had been surveying the isolated mountain ranges of New Guinea since 1964, when John Terborgh and he had made a preliminary survey to the eastern highlands of Papua New Guinea. After heavily working Papua New Guinea in the 1960s and 1970s, Diamond began to focus on the less-studied western (Indonesian) New Guinea—then called Irian Jaya. He was able to drop by helicopter onto a gravel river bar in the Foja Mountains in 1979. He struggled into the uplands and there found the lost bowerbird, which built its maypole bowers on forested ridge-crests.

Diamond returned in 1981, this time getting to 6,600 feet, and observed a number of new bird populations. He published a seminal paper in 1985 that described his discoveries, and his evocative description of the pristine environmental conditions in the Fojas set tropical biologists alight with interest. Here was a place that begged for follow-up in-depth surveys. Who knew what other undescribed plants and animals lurked in corners of this isolated mountain range?

While Diamond was ranging through the vast interior forests of the Foja Mountains, he also glimpsed a tree-kangaroo. What species was it? No red-colored tree-kangaroo had been recorded from Irian Jaya as of 1981, nor

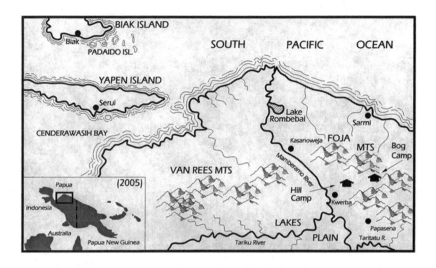

had any member of this lineage been described from the north coastal ranges of New Guinea east or west. It was only later, in 1993, that Tim Flannery described the Golden-mantled Tree-Kangaroo from a tiny remnant population in the eastern Torricelli Mountains of Papua New Guinea, far to the east of the Fojas. Might Diamond have observed this newly described creature, thought to be the rarest and most beautiful of the tree-kangaroos?

I first seriously discussed the Foja Mountains with colleagues when I was at the Smithsonian Institution in 1977. That summer I spent as a fellow in the research laboratory of Dillon Ripley, who had himself attempted to get into the Fojas in 1960, ascending about 60 miles up the Tor River from the north coast but being unable to get higher than the foothills. We both wanted badly to get there and knew nothing of Diamond's developing plans. Diamond announced the results of his two Foja field trips and his rediscovery of the bowerbird at a press conference at the National Geographic Society in 1982. The news only whetted our appetite.

In 1987 I finally got my first glimpse of these fabled mountains in a Cessna overflight. One crystal-clear morning I looked down on a huge expanse of undeveloped montane forest, broken only by the occasional land-slide. No villages, no logging tracks, no visible means of ready access. The mountainous interior of the Fojas looked a bit like the Great Smokies—verdant forested hills and rocky stream gorges. The relief, however, was considerable and the only access routes were a jumble of ridges—an explorer's nightmare. As we passed the high western summit, I glimpsed a tiny dry lakebed at about 5,000 feet. Eighteen years would pass before I would finally have a credible chance of choppering into that perfect upland clearing.

A lot happened between 1987 and 2005. During the decade I spent at the Smithsonian with Dillon Ripley, he and I worked to obtain permission from the Indonesian government for this expedition, and we raised funds for the field trip. Ironically, permission from the Indonesian government did not arrive until after I had moved from the Smithsonian to the Wildlife Conservation Society, and I read the approval letter with some bitterness, because at WCS my work was focused on Papua New Guinea. I filed the letter away wistfully.

In 1994 I started working for Conservation International. Jatna Supri-atna, Dedy Darnaedi, Adrian Forsyth, and I overflew the Fojas once more (Chapter 4) and initiated discussions about leading a Rapid Assessment expedition there. Permissions were slow in coming, and in 1995 both Forsyth and I departed CI. Once again the expedition plans were put on the shelf.

In the year 2000 I returned to work at Conservation International and Irian Jaya came back up on my radar screen. We started talking once more about the Foja Mountains as a target for a Rapid Assessment.

It was in early 2005 that I made a third overflight of the Fojas and saw the original dry lakebed on the ridge-top. Now it seemed to be simply a matter of waiting for the official government permissions to do the job. We got our letter of clearance from the Indonesian Institute of Science (LIPI) that autumn.

At this point I learned that the LIPI letter was just the start of the paper chase. We needed national-level clearances and permissions from immigration, forestry, interior, military, and the police, and then three additional provincial clearances. Most problematic was that these permits would not be obtainable until the entire field survey team had arrived in the country.

In mid-November the foreign team of four arrived in Jakarta: Steve Richards (herpetologist), Kristofer Helgen (mammalogist), Wayne Takeuchi (botanist), and me. A fifth lead fieldworker, Brother Henk van Mastrigt (entomologist), awaited us in Jayapura, where he lived. In addition, we had a counterpart team of six Indonesians who would meet us in Papua once we had completed our clearances. These included the scientists Yohanes Mogea, Yance de Fretes, Edy Sambas, and Burhan Tjaturadi as well as several graduate students from the two universities in Papua. We also had a strong field support team led by Nev Kemp and the Mamberamo CI field officers.

With the politically adept guidance of our CI office professional staff in Jakarta, plus a lot of advance legwork, the foreign team collected its Jakarta permits in four working days and moved on to Jayapura, the provincial capital of Papua—the new name for western New Guinea. An additional two working days in Jayapura produced the local permits—we had threaded the needle!

Our team had already transported the field supplies into our field base at Kwerba. On November 15 the scientific team started to shuttle into this lowland jungle staging site for the assault on the Fojas. Kwerba airstrip lay fifty-five minutes due west of Sentani airstrip, which served the capital city of Jayapura. We flew in the team and the remaining cargo on four Cessna charters.

I took the last flight. The day began cloudy and stormy. We arrived at the Tariku Airline hangar in Sentani in the predawn and loaded the tiny plane for departure (four passengers and field gear). We lifted off and headed west, only to hear from our pilot that the plane's alternator was acting up. The pilot circled back to the hangar to get the device repaired. Once the problem was fixed, we took off again and headed into the clouds, only to find that the alternator was malfunctioning yet again. Back to the hangar. Two hours later, with a third alternator in the craft, we took off into the building gloom bound for the Fojas. By this time the mountain summits were heavily shrouded, so our pilot was unable to give us a preview of the mountain summits we were hoping to ascend.

Banking in toward the Kwerba airstrip, we first crossed the Mamberamo River where its deep channel courses through the hills toward the north coast. We looked southward to the vast interior Lake Plain, the marvelous forested basin that in the 1990s had been targeted for damming to promote a giant industrial development scheme. Conserving that huge tract of tropical jungle was a major focus of Conservation International. We were hoping that by making a splash with our Foja Mountains exploration we might promote our field conservation efforts as well.

As the aircraft descended at near-stalling speed, we had plenty of time to fixate on the tiny landing ground ahead. The front of the airstrip sat atop a sheer cliff that dropped about a hundred feet down to the course of the Siri River. Given the minimal length of the strip, its poor surface conditions, and the heavy load in this small single-engine aircraft, the pilot needed to come in as low as possible without smacking head-on into the cliff, a feat that made most of us close our eyes on the last bit of the final approach.

The arrival of the plane drew the entire village out to the airstrip to greet us and help us with our loads of baggage. Kwerba was situated near the eastern bank of the Mamberamo, where the Foja foothills ranged down to the river. Kwerba, therefore, was the logical gateway into our mountain promised land. The village is typical of lowland jungle villages around New Guinea—small (about two hundred souls), surrounded by forest, and a product of the magnetic attraction of the airstrip and the organizing power of evangelical Christian missionaries. In other words, Kwerba was like so many of New Guinea's isolated villages with airstrips; it was a modern Western construct. Throughout this immense island, the villages served the convenience of governments and churches by organizing forest peoples into accessible and manageable units. Before the advent of airstrips, the indigenous people here lived scattered through the forest, in dispersed family compounds.

That said, the presence of Kwerba was useful to us as well. We needed informants, naturalists, and guides for our work. Here they were, brought together conveniently right next to the grassy airstrip, which they trimmed by hand every month or so. Our Conservation International field team had been working out of Kwerba for more than a year, so CI was familiar to the

locals. We had also established ourselves by constructing a spacious conservation center right in the village proper. This big hut made of local materials
would be our operational base for the expedition, where CI field staff would
be able to organize, and plan, and dine, and prepare for the big field trip.
Given that the CI support staff had to provision twenty-four fieldworkers
and assistants in two separate camps over three weeks, this daunting support
mission required considerable planning and logistic capacity.

The Rapid Assessment team's objective was to survey birds, mammals,
butterflies, plants, frogs, lizards, and snakes at three elevational zones: the
lowland forest (around Kwerba itself); the hill forest proper, reached by hiking eight hours northeast of Kwerba; and the montane forest, to be accessed
by getting a helicopter to drop us into the dry lakebed at 5,500 feet.

The Kwerba environs are a biologist's paradise—the confluence of New
Guinea's richest terrestrial environments. Verdant forested low hills descend
to a great river system and vast expanses of swamp forest. The physiographic
diversity creates abundant habitats for the flora and fauna. My first morning there, I awoke in my small tent to a rich morning chorus of birds of the
lowland rainforest. This chorus actually began in full darkness, and swelled
in volume as first light graded into dawn. The pioneers in this remarkable
phenomenon were the night callers. The little boobook owl and the Marbled
Frogmouth gave their last croaks of the evening, with the crepuscular whistles
of the Hook-billed and Shovel-billed Kingfishers mixing with the low hooting
of the Greater Black Coucal and bubbling jumble of the Variable Pitohui.
Dozens of species then chimed in to waken even the groggiest sleeping ornithologist. For me, it was the very best of wake-up calls.

Kwerba's forests pulsed with life: large fruit-bats, tiny insectivorous
bats, wallabies and tree kangaroos, scuttling forest rats, more than 120 species
of birds, Death Adders and Cat-eyed and Small-eyed Snakes, frogs large and
small, more than a hundred species of butterflies, and hundreds of jungle
plants including palms, pandans, mahoganies, figs, laurels, and the like made
for a biotic wonderland. The whoops and hoots and whistles of birds of

paradise came from all directions. I have never been to a jungle with more birds of paradise. In short order I listed the King, Lesser, Twelve-wired, Magnificent Riflebird, Pale-billed Sicklebill, and Glossy-mantled Manucode. More were in the hills and mountains above.

The prime objective of our expedition was to complete a first multi-disciplinary survey of the upland flora and fauna of the Foja Mountains. Achieving this objective would be a challenge for several reasons. First, lo-gistical problems involved obtaining a helicopter to ferry us up and down the mountain. Second, our field party was too large for a single camp. Third, the weather was poor, dominated by cloud and rain (the rainy season was just beginning). Fourth, we were in the middle of nowhere, and things have a way of going wrong when the support base is at a distance.

The helicopter issue was problematic. Jayapura, the provincial capital, had no helicopters. Instead, these machines were scattered about in local centers far from Kwerba that provided business related to mining and log-ging (primarily Nabire and Timika). Helicopters are fragile and expensive machines, and we had only so much money to invest in rental. HeliMission, a nonprofit Christian missionary assistance group, agreed to allot us a single day in and a single day out, all the charter time they could spare us. The days they allotted to us, November 22 and December 7, gave me pause. The first was the anniversary of John Kennedy's assassination and the second was Pearl Harbor Day—not a propitious pair of dates for an American. Given the uncertain weather and conditions in western New Guinea, I was not brimming with optimism.

We had to divide our big group into a "mountain" team and a "hill" team. Selecting teams was politically sensitive and had to be finessed so as not to create friction or jealousy. It was first agreed that CI staffers Yance de Fretes and Neville Kemp would lead the hill forest team, which would hike into the hills and establish a camp on the nearest high ridge at about 3,000 feet elevation. The mountain team would chopper into the upland lakebed at around 5,500 feet. We left the final decision on personnel assignments for after our initial field reconnaissance of the Foja hill forest.

In order to establish the hill camp, Nev Kemp and I carried out a rapid

walking tour of the hill forests northeast of Kwerba. We conferred at length
with our Kwerba hosts, who assured us that we could climb an adjacent ridge
and gain sufficient altitude to see some of the special wildlife of the Fojas.
We took them at their word.

The reconnaissance was an education. We thrashed around in the
rugged ridge country for three days, then dragged ourselves back toward
Kwerba, exhausted and considerably worse for wear. My legs were a mass of
briar cuts and scratches. One spot on my shin was already badly infected with
a tropical ulcer. I moved at this point with the aid of a walking stick. I cursed
each time my foot caught on a tangle or when I slipped in deep mud. Every
muscle and joint ached. Stumbling into Kwerba near nightfall, I stripped off
my filthy clothes and took a long soak in the stream and fell into bed for a
long night's sleep. The Fojas were not going to give up their secrets without
a struggle. We did manage to select a site, Hotice, for the hill camp, but it was
not in any way ideal—too low, too rugged, too difficult of access.

We now understood how little the local Kwerba people knew of the
interior Foja Mountains. There were no major walking tracks, few hunting
traces, and little real understanding of the backcountry by even the best hunters in Kwerba. These people had forgotten their past relationship with the
mountain range. Yes, they had many stories of the mountains and yes, they
knew the wildlife of the mountains, but they themselves had become seden-
tary. Hunting conditions were so favorable near the village that there was
no reason to range farther. Cassowaries, pigs, wallabies, and tree-kangaroos
existed in abundance within an hour's walk from the village. Crocodiles and
various fish inhabited the Siri and other rivers. This was the land of plenty.
As a result, the local naturalists did *not* know the ridges leading up into the
clouds the way their grandfathers did.

We had hopes of cutting a ready access trail to 3,000 feet elevation a couple
of days' walk from Kwerba. Now that hope was fading. Even though the hill
team could do its work in Hotice, it would probably not be able to get to even
the lowest reaches of the montane flora and fauna. That mandate would fall to

This helicopter, our only transport, after four trips was fogged in at our clearing
in the interior uplands of the Foja Mountains.

the upland team. And it was the interior uplands above 3,500 feet that would
hold most of the biological secrets of the Foja Mountains. Thus the success
of our expedition depended on the helicopter. Would it come? Would the
weather permit us access to the mountain interior? Would we be stuck in the
lowlands, with the bowerbird and other mysteries just out of our grasp?

November 22 was our helicopter pickup day. In anticipation, we were
all up before dawn, packing tents, weighing baggage, nervously pacing, and
listening for the distinctive cadenced *d-d-d-d-d-d-d-d* sound coming in from
the south. We hoped for an early arrival, before the clouds began to build
over the upland interior.

After what seemed like an eternity, the unmistakable sound of rotor
blades brought a shout to move cargo out to the airstrip. Brian, the pilot,
refueled, and we packed the craft to bursting with four passengers and a lot of
bulky but lightweight baggage. With a full gas tank, the pilot needed to travel
light on his first sortie up the hill, especially because he was heading into an
unknown mountain zone that could require some agility. At this point, the
only guidance the pilot had was my set of GPS (Global Positioning System)
coordinates and an elevation.

Because of my overflight experience, I was selected to go on the first of our five planned sorties. We lifted off gingerly and headed northeastward, the pilot following the direction provided by his dashboard GPS. Before long we could see vast pillowy clouds resting atop the highest summits. The helicopter continued to climb, following a long ridgeline, and soon the pilot began to probe the edges of the cloud mass.

There would be no direct flight to the lakebed landing zone this morning. After about twenty minutes aloft, the pilot informed me on the headset that we were now within a mile of the lakebed but that clouds blocked the final approach. We then began to circle the area, looking for a break in the clouds. No luck . . . The pilot then took us northwestward toward the west summit, just to kill time, I suppose. In desperation I started thinking about a fallback plan. Could we be dropped into one of the open-stream valleys below the clouds? Because of the abundant rocky landslips and sandbars, many of the larger streambeds were quite open and would accommodate a helicopter. But we would sacrifice a lot of elevation. No, it would be much better to get into that dry lakebed atop the high ridge.

Here we were, defying gravity, circling above the mountain forest. I wondered what the pilot was planning to do. How long could we temporize up here in the middle of nowhere? We did scout a few other boggy openings on the summit ridge. These offered some possibilities, but they were littered with small saplings.

Thank heaven for the GPS! That little miracle of a device could help the pilot keep his eye on the target destination even though clouds obscured it from view. Though I was fast losing hope, apparently the GPS allowed the pilot to be patient, understanding that with passage of time and a crack in the clouds, the mission could be achieved.

As we drifted back to the southeast, the clouds parted and the boggy lakebed was in clear view. I finally exhaled. We dropped rapidly toward the landing zone, and the pilot debated where to touch down. Was the area of thick marsh grass firm? Or was the zone of bare earth better? The pilot chose the bare earth, and the ground held as he ever so gently reduced the rotors' lift. Within seconds we were clambering out of the noisy machine with the blades still whirling above us, hunched over, clawing at the mass of baggage, lugging it to the grassy part of the lakebed just out of reach of the deadly blades. Four minutes later the machine lifted off and disappeared over the brow of the ridge. The sound receded, and the four of us, stunned, stood in our *Lost World*—the place we had wanted to visit for so long.

It was late morning, and the sun shone around the swirl of misty cloud that was rising up to this high pass. It was quiet except for a *Melidectes* honey-eater gurgling in the forest, and a few other birdsongs I did not recognize. We stared about in wonder. I decided that we needed to prepare a landing ground. The lakebed was in fact a sphagnum bog interlaced with tiny chan-nels of flowing water. Some areas were quaking and waterlogged, others were firm. It was treacherous. We checked out various patches and decided the pilot's original instincts were best. We marked out a landing zone with pink flagging tape and started cutting a path into the adjacent forest. We needed to select a place for a campsite before the afternoon rains.

The chopper returned three more times in forty-minute intervals. Each time, people and baggage and supplies were disgorged in a swirl of wind and echoing rotor sound. Each return produced a heart-pounding scene where people and cargo moved wildly to and fro, accompanied by the deafening sound of the great machine that made it difficult to communicate instruc-tions. The sound was magnified by the basin shape of the boggy lakebed. There is nothing like a hovering helicopter to break the silence of the pristine wilderness.

We selected a flat knoll in the forest as our campsite, and the Kwerba men began to clear the area and select saplings as building timber. The fourth helicopter run was a close call. The low cloud, which arrived up the ridges from the humid lowlands, was closing in on our new forest home when the pilot punched through and delivered his cargo. The weather ended all hope of a fifth run. We had our full complement of fieldworkers—twelve in all—but we were missing 400 pounds of equipment and supplies. We were socked in, with dark clouds and rain on and off. By the end of the day we knew our helicopter would not be back. Would it be able to return another day to deliver our critical supplies?

During our first hour up top we were nervously elated that we finally had made it to this promised land. We had surmounted the many hurdles and had defied the odds and had made it into the Fojas! Within minutes, each of the scientists was off inspecting some corner of the bog. Kris Helgen was following the abundant animal tracks through the marsh grass, looking for wallabies. Wayne Takeuchi was hunting for shrubs in flower along the border of the bog. Steve Richards was listening to the tinny sounds of frogs calling from the grass. Brother Henk van Mastrigt was eyeing little white butterflies that were flying just out of reach. I craned my neck in search of birds moving along the border of the forest.

I was standing in an old lakebed atop a flat ridge high in the interior of the Foja Mountains. Most of the south side of the lake had filled in and now was a walkable sphagnum bog. Patches of open water lay on the north side. Thickets of rhododendron and other shrubs formed a wall around the bog's border, and tall montane forest rose up beyond. With clouds obscuring the sun, the scene was stark and wild. I was far, far from home.

When we all met for a late lunch on our first day in the mountains, several of the party mentioned their encounters with a weird bird with dangling orange wattles. I initially thought they were describing a Wattled Ploughbill, which would have been a surprise in this north coastal range. But the bird in question apparently also had a bright red face. I got mixed descriptions from the various informants, so I was confused but intrigued. (One of the Kwerba men said its face "looked like a chicken.")

I didn't actually see the bird for myself for another five days. When I did encounter it, I recognized the reason for all the amazement. My colleagues were describing a species new to science. I had not imagined we would so quickly encounter such a find. The last new full species described from the island of New Guinea was the Snow Mountain Robin, discovered by Austin L. Rand in September 1939, just before the beginning of World War II.

This Wattled Smoky Honeyeater was much like one of its more widespread cousins. In fact, I studiously ignored the first few of the new species that I encountered at the edge of the bog, because I assumed they were Common Smoky Honeyeater—an annoyingly ever-present species in the montane forest of New Guinea's central cordillera, the bird that appears in your binoculars every time you are looking for something rare and desirable. So I shrugged off these smoky honeyeaters and simply included them in my day lists as "*Melipotes,*" not giving them another thought. My lesson that it pays to look closely at *every* bird when in a new environment!

This new bird's bare face patch was red-orange, not orange-yellow. The face patch was larger and more irregularly shaped than that of other honey-eaters. Incredibly, the bottom of each face patch ended in a free-dangling wattle of skin the same color as the face, like no other honeyeater in all of New Guinea. We had a new species!

Then there was the "lost" bird of paradise. Earlier in this chapter I wrote of Otto Kleinschmidt's description of the mysterious *Parotia ber-lepschi.* This form was subsumed into another species by Ernst Mayr in his influential *List of New Guinea Birds,* published in 1941. With time this bird was essentially forgotten by the ornithological world in a way that the Golden-fronted Bowerbird was not (thus the difference in visibility of a "spe-cies" versus a "subspecies"—which can be an arbitrary decision in birds as poorly known as these).

On our second day, a male and a female *Parotia berlepschi* appeared at the edge of Bog Camp and put on a display that mesmerized those of us in the camp. We stood in awe as the male romped about in some low saplings around our entrance trail, flicking his wings and white flank plumes, and whistling his sweet two-note song for the female. He then dropped to the ground in the middle of the path, hopping to and fro in full display. I was too spellbound to go get my camera. Too bad, it would have been a stunning series of photographs.

On subsequent days this male would traipse through the canopy of the forest at the edge of the bog, singing and moving about conspicuously and announcing himself to the world. The day before we departed, I was able to trace the male en route to his terrestrial dance ground in a thicket in the forest. Here was his secret trysting site, where he would bring females for mating. I was never privileged to see that love-interaction, though I did see the male dance and prance and pose there with a ball of moss in his bill, perched on a branch above his dance floor. At one point, an immature male came and watched from the overhanging branch as the adult male did his balletic routine on the ground.

The minute I saw Berlepsch's Six-wired Bird of Paradise, I could tell it was distinct from Carola's Six-wired Bird of Paradise. Unlike Carola's, Berlepsch's was essentially a black bird with white flank plumes; its voice was distinct as well. Here was yet another bird species unique to the Foja Mountains. Add this to the wonderful Golden-fronted Bowerbird, redis-covered by Diamond here in the Fojas, and the Wattled Smoky Honeyeater, and suddenly the Foja Mountains became an important area of endemism for birds in New Guinea. Steve Richards was later to assemble a collection of twenty new species of frogs from the Fojas. Wayne Takeuchi identified several new plant species (the determinations are not yet complete). Brother Henk found five new butterflies, and Yohanes Mogea discovered a half-dozen species of new palms. We expect that Kris Helgen will eventually describe a handful of new mammals, as well.

This was indeed a Lost World, a land that in many ways had been overlooked and left behind. Even the village landowners had little knowledge of this lost forest so many days' walk from any village. They had heard stories told by aging relatives around the campfire, but had no firsthand experience. Our local Kwerba and Papasena hosts were seeing all sorts of novelties during their first visit to the mountain.

We were at Bog Camp with a group of six local landowners, including the two senior leaders from Kwerba and Papasena (Isak and Timothy). Re-markably, their knowledge had been gained from informal teaching through stories told by elders, relatives, and loved ones: Isak and Timothy, although they had never seen this site and had only penetrated the verges of this isolated range on hunting trips, knew all about the wildlife up here. We had discovered this the preceding week back in Kwerba, when we had queried them on the mammals and birdlife, using illustrated books to focus the dis-cussion. Without prompting, the men had eagerly pointed out obscure and little-known mountain species and indicated that these indeed lived in the Foja Mountains—species such as the Golden-mantled Tree-Kangaroo, the

Golden-fronted Bowerbird, and Berlepsch's Six-wired Bird of Paradise (for this last they had pointed to Carola's and said, "The Foja species looks a bit like this but is distinct").

The oral tradition clearly thrived in this corner of Papua. Ancestors had spent time in the mountain interior and had shared their knowledge of the mountain fauna with their offspring and grandchildren. The two village elders, camped up top with us, were as excited as we were whenever the team located one of these "missing" creatures.

The Fojas were special because they were pristine, untouched, uncropped, unvisited. The island of New Guinea is cloaked in forest, but few of New Guinea's forest tracts have never been visited by the traditional landowners. Even isolated mountain forests receive visitors regularly, especially hunters searching for wild game, or travelers heading cross-country to a distant destination, or shamans in search of special plants. People use the forest in ways we can scarcely imagine in our urbanized and commercialized Western world. Witness the okari-nut hunters that I encountered in the Lakekamu Basin (Chapter 7).

The Fojas are different, however. The human population in the area is so small and so scattered and so confined to the edges of this vast world, that the core forest block apparently is today *entirely* free of human influence. In our two weeks ranging out in all directions from Bog Camp, our team of twelve never encountered any evidence of humankind present or past. It was a wild land given over to wildlife.

Places such as these are so very rare on earth that it is difficult to appreciate their significance. They are, in fact, the remaining "natural earth prototypes" that have escaped being colonized and modified by humans over the last several millennia. In the early 1980s, Diamond wrote about the significance of pristine worlds like the Fojas. We now saw firsthand what he was talking about. We quickly came to realize the global significance of such a precious environment that is free of cats, black rats, myna birds, starlings, House Sparrows, strip malls, paved roads, trails, sport-utility vehicles.

Our fifteen days in the Lost World went by with remarkable speed. Each day we worked from before dawn until after dark. Some of us rose before five to record birdsongs. Others went out after dark to search for frogs until a few hours before dawn. Hunting parties launched out in the dark, in search of unknown mammals higher up the ridges. We wandered far and wide trying to unlock the Fojas' secrets.

Some aspects of our work were easy (encountering the new smoky honeyeater and the lost six-wired bird of paradise). But there were tough times as well. Because of persistent rain and fog, our entomologist, Brother Henk, could hunt his butterflies for only about an hour a day. We were cloud free only in the early morning and after dark. Shortly after the sun rose, the clouds billowed up from the lowlands and filled this pass between the mountains. We had rain for some part of every single day; most days it rained or drizzled four or five times. Heavy mist settled on the forest for hours at a time. The ground was saturated, and the walking paths around camp became quagmires. I never once wore on my feet anything but my high rubber boots, in the camp or out in the forest. The camp itself became a mucky mire, because the big plastic roof flies all drained inward toward the middle of the camp. We had no place to bathe without encountering mud. The only time we felt clean was when we were asleep in our sleeping bags.

One morning, after a night's heavy rain, we awoke to find that the bog had become a lake again. My green tent, situated on a small hillock at the edge of the lakebed, was entirely surrounded by standing water and I had to evacuate to the forest up the hill. The helicopter landing ground was covered with a foot of water. Our camp trails led down to the lake edge. What would we do if these conditions persisted until the morning the helicopter came back to get us?

In spite of all the bother, it *was* a paradise for us. Our spirits soared with the joy of being in this incredible place, so far from everything, surrounded by wildlife and trackless forest.

⬥⬥⬥

My days on the mountain tended to follow a pattern. I rose in the cool predawn hours and tromped out to the bog to make dawn birdsong recordings. I then opened the mist-nets. I visited a blind to observe the display of the

Golden-fronted Bowerbird or else I conducted a walking bird census. I alternated checking the nets, measuring and recording the birds netted, and doing midday and afternoon censuses. I also surveyed and marked new walking tracks. What I found myself doing, mainly, was walking. I must have walked ten hours each and every day. By eight at night I was physically spent. I had supper and crawled into my mountain tent, stripped off my mud-encrusted clothing, and was asleep in minutes, lulled by the sounds of the Papuan Boobook Owl and the Feline Owlet-nightjar.

Everybody found something new. We were on the field expedition that each one of us had dreamed about. There simply was not enough time to catalog all that we were encountering. We could see that the fifteen days we were spending here at our mountain camp would allow us to get only part of the job done. This was especially true for the plants and the mammals, two important components of the biota. The surveying of both groups suffered because a lot of the necessary equipment and traps had been left behind. (HeliMission never did come back with that 400 pounds of supplies.)

Kris and Wayne labored on, in spite of this setback. Kris was heavily dependent on the Kwerba men to hunt mammals at night. Once they came back with a Long-beaked Echidna, one of the rarest and weirdest mammals on earth. Weighing about 30 pounds, with protective spiny quills on its back like a porcupine and a long snout used to probe the ground for earthworms, the Long-beaked Echidna is one of the egg-laying monotremes—the most primitive of living mammals (a relative of the platypus). Although it also sports a poison claw on each hind foot, this creature is slow and harmless. Our individual was found by Steve Richards right on the ridge trail. Richards put the echidna in a sack and the men happily carried it back to camp. We encountered three individuals of this rare species in a five-day period.

The most wonderful mammal we encountered was Jared's mysterious red tree-kangaroo, which turned out to be the Golden-mantled, described by Tim Flannery in 1993. The Kwerba men encountered this wary species several times in the forest at night, but had great difficulty keeping up with it. They finally brought an individual in to camp early one predawn, and the shouting that followed woke us all. This golden-and-red-furred beast sported powerful clawed fore and hind paws for climbing trees and a yard-long tail that helped it balance when climbing. This was New Guinea's analog to a Red Colobus Monkey, a familiar creature in central African forests.

Wayne collected more than seven hundred different plant samples. Some were so bizarre that he immediately knew they were new to science. Many others were different from the species he knew. It will take him many months to work out species determinations for this prodigious collection. Given that few plant species are flowering at any particular time in these tropical forests, Wayne was able to survey only a small sample of the flora. Additional visits will be necessary.

❧❧

Our departure day, December 7, came all too quickly. With some regret, we rose early to pack. All of us felt upbeat about what we had achieved, and once we were inured to the inevitable departure, we began thinking about the luxuries we would regain in the Kwerba lowlands: bright sunny days, bathing, a mud-free camp, lazing about in sandals and shorts.

The morning broke clear and promising, but one hour passed into the next with no sound of rotor blades. I rang HeliMission on the satellite phone and learned that the helicopter was busy on another charter and would not arrive until the afternoon. This we immediately knew was bad news. We had not had a cloud-free afternoon in fourteen days, and we did not expect this day to be any different.

We had been told by the HeliMission people that only one day was available to get us out. Now we were hearing that the day no longer belonged to us. Would we be spending Christmas up here in Bog Camp?

Noon came and went, and we listlessly went through the motions. Food, repacking, checking the details of our exit plan. Exit plan . . . What would we be facing if some of us got out and others did not? Who got to depart first, and who would be last? What equipment and supplies would we leave, and what would have to accompany us? How could we avoid over-loading the helicopter? How could we make certain that those who were stranded up top had sufficient supplies to last until a helicopter could come back to get them? We were nervous as cats, visualizing all sorts of unplanned and unfortunate outcomes.

At 4 P.M. the sound of rotors cut through the mists that surrounded the bog. We could not see the craft. It disappeared off to the northwest. Minutes later the sound returned. Then faded off. No opening in the clouds. All of a sudden the whirlybird came rushing in from above and dropped right down onto the original helipad, which we had improved with timbers and new markers. We quickly loaded up people and cargo; pilot Tom lifted the bird and headed out . . . into a great gray wall of cloud. We probed here and there, but the pilot could find no break in the wall. After five minutes of frustration we dropped back to the landing zone and waited five minutes, the rotors whirling. We rose again and tried our luck. Again there was no exit.

We dropped back down to the bog and powered down. Everyone piled out and headed to camp for a cup of tea and a biscuit. We had to wait out the cloud and mist, but time was short. The sun, wherever it was, was sinking.

I kept walking down to the bog clearing, looking for some break in the weather. A bit before five, we saw some indication of hope and piled back in. Up we lifted, and Tom picked his way through cracks in the clouds. We were over the ridge and heading down. But then, every direction we turned brought obscuring walls of cloud. Tom zigged and zagged, and we found ourselves in a maze of mountain ridges and low cloud. Not good. It was not certain we could make our way back to the bog, yet we couldn't see a way down to Kwerba.

Tom drifted lower and lower into a forested ravine, looking for passage. The clouds formed a solid ceiling above us. I searched for a place to put down if necessary. Tom radioed his base to say he had to return to Bog Camp. Turning to do so, he saw a ray of light to the left and made for it. We squeaked through into another small clearing in the clouds, and slowly but surely began to extract ourselves from our frightening predicament. My fingers began to loosen their grip on the seat, as I started to believe that we would indeed make it to Kwerba that day. For a long moment there, I had been steeling myself for the unpleasant experience of spending the night in a rocky stream gorge after an emergency landing.

Suddenly the forest below exploded with birds—the snowy white of

Sulphur-crested Cockatoos, the black-and-red of Vulturine Parrots, and the black-and-white of Blyth's Hornbills. It was as if they were rising up to show us the way to Kwerba.

The most remarkable aspect of this trip was that once we had escaped the clouds, we found ourselves with a free shot at Kwerba and we rode over ridge after ridge getting there. I looked down at that forest wilderness, all uninhabited, and rather than marveling at how wonderful it was, I was thinking about how horribly long it would have taken to hike from our mountain bog camp back down to the Kwerba lowlands. Each time I thought we were finally there, more forest passed below with no sign of the airstrip.

Finally we saw a garden, then another, then the grass strip! As the chopper gently lowered to the landing ground, Kwerba felt like home sweet home.

Even though some of us got to bathe in the lovely stream pool that night, others remained stranded up top in the mists. It was not until the next morning that our pilot extracted the rest of the team and cargo from the mountain bog camp. By midday the groups from the hill camp and the mountain camp held a joyous reunion at our research station in Kwerba. People were bubbling over with excitement, sharing discoveries and stories. We were all back from the forest, safe and healthy. We had done it! We had explored the Fojas!

After two days in Kwerba—washing, drying, mending, labeling, organizing—we found ourselves in a milling crowd of village people, young and old. It was our farewell party, and every human in the Kwerba area was there to celebrate. We pooled our food resources and laid out a feast. Our group donated a lot of exotic items that were a special treat to the Kwerba residents. Before eating, a pastor said a benediction, and many of us made speeches, praising one another and the success of the effort. The speeches came to an end with much hugging, and then we had a free-for-all with the abundant food. Millions of tiny gnats joined the party, attracted by the bright lanterns that had been hung. Before long we were all drifting back to our tents, exhausted by the excitement. We were drained from the three weeks without letup, and we were ready to pack up and head home for the Christmas season that was fast approaching. We had spent Thanksgiving up in the bog, and we wanted to enjoy Christmas with our families.

⤙⤚

There really is nothing quite like finding a new species in nature. It is a prime motivation that keeps many field naturalists young at heart. For some groups (frogs, most insect groups), finding new species in tropical forests is almost commonplace. For other groups (butterflies, birds, mammals) the hunt is much more difficult, mainly because of the excellent work done by our predecessors. They left only a few discoveries for this current generation to make. In some ways, that fact makes the lure of the search even greater.

The real wonders of the Foja Mountains were the new frogs and plants. Steve Richards and Wayne Takeuchi brought in species unlike anything they had ever seen or studied. Though they were used to locating new species wherever they ranged on this enormous island, they were stunned by what they found.

The discovery of forty new species in the Fojas touched a chord with people around the world. It was the single issue that resonated most. The phenomenon of discovery awoke a dormant passion in many individuals, who emailed us from far and wide. People have a primal fascination with something new in the world, with something unknown and uncharacterized, with wild creatures that eluded the naturalists of the world until the present moment.

The issue of discovery is allied to the issue of unknown and unvisited environments, also of strong interest to the reading public in the twenty-first century. Most people today think of the world as mapped and scrutinized and easily visited on Google Earth. In fact, there exist little-known corners of the globe that have been, in effect, "forgotten" and that are pristine or nearly pristine. These are waiting to be "discovered" or "rediscovered." These are the wonderful lost worlds that will, as they are visited and announced to the world, excite the imagination of children and adults alike, giving hope for the future and reminding all of us of just how little known some parts of our world are. We biologists have our work cut out for us and should not be in a hurry to start expensive investment in documenting the life forms of Mars and the trans–solar system planets. We have an unknown earth right here in isolated mountain ranges like the Fojas (and the Wandammen and Kumawa and Fakfak mountains), as well as in the ocean deeps and atop the many constellations of seamounts that are scattered within the oceans.

⋦⋧

A recounting of the origins of the Foja Mountains can help us understand why endemic species live up there. The first hint is the rock one finds atop the summits—dark gray deepwater shales. The piece of earth fragment on which the Fojas sit was under the sea and disjunct from the New Guinea mainland two million years ago. One million years ago, the land that became the Foja Mountains was a newly conjoined patch of lowlands that had recently rafted into New Guinea's north coast. This tectonic collision started the compression that led to the uplift of that former patch of sea floor into what was to become the Foja Mountains. We can surmise that the highland habitats that support all those Foja endemic species did not come into being until perhaps a half million years ago. It presumably took some time for the montane plantlife to colonize this new "island" of mountain habitat that had risen from the sea not long before. These montane plants dispersed from the central range to the south of the Mamberamo Basin. They must have been followed by the arrival of predecessor species of birds, mammals, frogs, and butterflies—which, in isolation, differentiated into new-minted endemic species.

In effect, the Foja Mountains are a new "island" of isolated montane habitat into which a whole biota has colonized, adapted, and in some cases differentiated. How did those montane species actually get to the isolated summits of the Fojas? Our best guess is that the process was aided by Pleistocene climate change, well after the summits had been uplifted. During a period of global cooling, we believe that montane plants and animals shifted their elevational ranges down into the lowlands. These downward shifts could have exceeded 3,000 feet of elevation over time (and for millennia at a time). Thus species that lived at 3,000 feet during "normal" times moved down to sea level during the cold period. Montane species flooded into lowland areas. When climate conditions ameliorated and temperatures rose, these displaced populations began to rise back up the mountainside. Presumably populations living near the base of the Foja Mountains ascended into the Foja uplands, providing a "seeding" of cold-loving forms. This climate-driven process required no long-distance dispersal from the distant central range to the south. The species instead crept up into the Foja Mountains, following the cool climate as it rose up the hill.

The Foja Mountains are a textbook example of the interplay between plate tectonics, mountain building, climatic fluctuation, and the evolution of a new montane biota. Additional study will give us more data to fine-tune our knowledge on the timetable of speciation and the nature of community assembly in newly evolved habitats.

❧❦

Earth history includes both the ebb and flow of species. I learned about mammalian extinction in the forests of New Guinea on this expedition thanks to an informal field education by Kris Helgen and the examples of the Golden-mantled Tree-Kangaroo and the Long-beaked Echidna. The mammal story was quite different from the bird story.

My own study of the New Guinea bird fauna taught me that few species of forest-dwelling birds in New Guinea are threatened by hunting, as long as large expanses of forest remain in place. Thus if we conserve blocks of natural forest, we believe that no birds will go extinct.

What I learned from Kris is that a number of New Guinea mammals have suffered extensive local extirpation and that some more vulnerable forms have disappeared altogether in the recent past, the results of nothing more than chronic traditional hunting. Apparently an early extinction pulse of large-bodied mammals had occurred in the prehistoric era—large wallabies and marsupial sloths (diprotodonts). More surprisingly, ongoing pressures currently seem to be threatening certain of New Guinea's largest native mammals.

Consider the echidna and tree-kangaroo. The 30-pound Long-beaked Echidna is slow moving and much appreciated by hunters all over New Guinea for its tasty meat. Once widespread there, the range of this echidna

Mammalogist Kris Helgen holding a Golden-mantled
Tree-Kangaroo, a rare species here recorded for the
first time in Indonesia.

is now patchy. It is found only in places with few people and minimal hunt-ing pressure, mainly in the highest mountains farthest from the villages. The story of the tree-kangaroo is more extreme. Tim Flannery had heard about this species in Papua New Guinea's North Coastal Range. Tribesmen from several areas had reported a red tree-kangaroo that had once been a favor-ite food but now was gone. After several years of searching, a member of Flannery's team located a single individual on an isolated mountain at the easternmost terminus of the range. By all accounts, the species had been hunted out from more than 95 percent of its known range.

The lesson here is that mammals such as echidnas and tree-kangaroos are vulnerable to extinction through subsistence hunting in continuous for-est lands. The secret is canine: men hunting with dogs can find every last individual, because of the dogs' keen sense of smell. So it is not just a matter of conserving forests, but of conserving forests free of men hunting with dogs that is necessary to protect the tastier mammals.

Discovery of the Golden-mantled Tree-Kangaroo in the Fojas made us all happy, because we knew that we had a vast forest tract free of hunting or any other human disturbance. Here the critically endangered tree-kangaroo and the bizarre echidna could live in safety.

※※

In the untouched forests of the Foja Mountains, Dwarf Cassowaries with no knowledge of human hunters search each day for fruits on the forest floor. New Guinea Harpy Eagles perch like gray ghosts in the mossy canopy, waiting for unsuspecting giant rats. The magnificent, sabre-tailed Black Sicklebills rattle their advertisement songs at dawn. In this superbly peaceful environment, I could walk quietly up to the display bower of the Golden-fronted Bowerbird, sit on a fallen log, and watch the male do his thing at his "love tower" of sticks and moss. Ignoring my presence, the female would sneak in to an adjacent perch, sending the crested male into a frenzy of excitement. He collected a tiny blue fruit in his beak and posed stiffly, beak upward, facing the moss-and-stick tower, muttering weird sounds under his breath, expanding his golden crest to prove his biological worth to the choosy visitor. What a pleasure to know that this beautiful and little-known bird species can continue to follow its annual cycles of construction and passion and reproduction in a rainforest that stands apart, isolated from the heavy hand of humankind.

Have we now robbed this Lost World of its protective and seductive mystery? Not exactly . . . We still talk about the ever-elusive black tree-kangaroo that lurks in the highest forests of the Fojas. Whether or not it was actually seen by the local naturalists on our trip is hard to determine. Other hunters near Kwerba brought us the skull of a tree-kangaroo from the Fojas that is larger than any living *Dendrolagus*. What might this be? It is difficult to say without more data. Then there are the detailed reports of a golden bowerbird of a kind that does not quite match the illustration in our field guide . . . These are all mysteries to be explored by future expeditions into the jungle. Lost worlds will continue to haunt our dreams.

Epilogue

My field trip to the Upper Watut of Papua New Guinea taught me that local people have much to teach the Western world about the rainforest and how it works. Moreover, what I learned in the Lakekamu Basin was that local people need to remain the stewards of their forests. They may require some help from the outside, but they must remain the leaders of any process that promotes conservation and sustainable development.

My trip to Panama taught me the obvious: that the Neotropics are the heartland of the earth's greatest concentration of biodiversity. Even giant New Guinea cannot compete with tiny Panama in terms of the vast numbers of plants and animals. Further, I learned that tropical forests have physical similarities across continents and should be managed at the global level to protect critical global environmental processes (the hydrological cycles, for instance). Consider the Hooded Pitohui and the poison dart frog. The life histories of these two disparate life forms are closely linked across space and time. Before long, I suspect we will discover that some fungus or plant or microbe inhabiting both South America and New Guinea manufactures the potent chemical that the bird and frog assimilate and "wear" as armor to protect themselves from predators. What additional plant toxins does this unknown organism produce? Might any be useful to science and medicine?

The trip to Irian Jaya taught me about the world of great mining companies and their impacts both locally and more generally. Collusion with the military and with venal government officials often lead mining companies down a dark path. The inability to address the needs of rural indigenous stakeholders leads to additional negative impacts. The mandate of "providing shareholder value" is a clear directive to international mining operations to get the gold, but at what long-term cost, locally? Governments of developing nations rarely can resist the lure of mega cash flow from mineral operations that are paid for entirely by outside interests. Yet true development comes from the doing—from learning the process and developing local capacity. Strength comes from effort, from trial and error. Dependence on a royalty check can never lead to smart development or sustained growth—of an economy or a society.

≈☙☙

What can be learned from our sojourns in India? The perils of scientific imperialism are worth noting. Close collaboration among researchers is a must that benefits everyone and reduces the potential for any claim of scientific imperialism. This issue will persist unless feelings of inequity between scientists from developed and developing countries are dispelled.

Scientists tend to be naive and self-centered. Everything is centered on the fieldwork. Researchers, in fact, need to be much more in tune with the local issues where they work. I learned this the hard way in the Lakekamu Basin, and also in Arunachal Pradesh, where suspicion of the true nature of our activities was high. Collaboration means mutual understanding, not just sharing the same tent camp. The same caveat would apply to our Foja Mountains expedition as well. Sharing and sensitivity are the watchwords of effective international scientific research collaboration.

Madagascar and Cebu gave a look into the not-so-bright future of the world's rainforests, should our global resource–exploitationist habits continue unabated on this planet. Why can't more national leaders wake up to the costs of failing to plan for the future? Strong economies are based on long-term thinking and strong environmental principles. A century from now we will all need fertile soil, clean drinking water, and abundant lumber. There is no high-tech bailout for ill-conceived resource stewardship.

What of Africa? What do we learn from the troubles in Ivory Coast? Certainly, the leadership of any developing nation would benefit from knowing how to avoid the political meltdown that ravaged Ivory Coast. The temptation is always for some strongman to take over and abolish the parliamentary system (because, he will say, democracy is messy and inefficient). That might provide a temporary fix, but it will lead to long-term chaos, as we have seen. Africa also shows that local and national politics are critically important to the environment. Appropriate environmental decisionmaking rarely comes as a by-product of poor governance. Governance is important and can not be sidestepped, even though some environmentalists would like to think it possible to implement long-term conservation in the absence of effective government. That simply is not true.

Everything matters when it comes to conserving tropical forests and biodiversity. The "biodiversity" sector can not be isolated and conserved in its own little bubble. We conservationists have to ensure that all the threads tie together: good governance, sensible economics, strong planning, enforcement, engaged local stewardship, and, yes, creation and management of protected areas to preserve the most precious places on earth.

Bibliography

Ali, Sálim. 1986. *The Fall of a Sparrow.* Oxford University Press, New Delhi.

Ali, Sálim, and S. Dillon Ripley. 1983. *A Pictorial Guide to the Birds of the Indian Subcontinent.* Oxford University Press, Delhi.

Beehler, Bruce M. 1989. The birds of paradise. *Scientific American.* December 1989: 117–123.

Beehler, Bruce M. 1991. *A Naturalist in New Guinea.* University of Texas Press, Austin.

Beehler, Bruce M., K. S. R. Krishna Raju, and Shahid Ali. 1987. Avian use of man-disturbed forest habitats in the Eastern Ghats, India. *Ibis* 129: 197–211.

Beehler, Bruce M., Roger F. Pasquier, and Warren B. King. 2002. In memoriam: S. Dillon Ripley, 1913–2001. *Auk* 119: 110–113.

Beehler, Bruce M., Thane K. Pratt, and Dale A. Zimmerman. 1986. *Birds of New Guinea.* Princeton University Press, Princeton.

Collar, Nigel. 1998. Extinction by assumption: or, The Romeo error on Cebu. *Oryx* 32(4): 239–244.

D'Albertis, Luigi M. 1880. *New Guinea: What I Did and What I Saw.* Sampson Low, Marston, Searle and Rivington, London.

Diamond, Jared M. 1972. *Avifauna of the Eastern Highlands of New Guinea.* Nuttall Ornithological Club, Cambridge, Mass.

Dinerstein, Eric. 2005. *Tigerland.* Island Press, Washington, D.C.

Dobson, Andrew P. 1996. *Conservation and Biodiversity.* W. H. Freeman, New York.

Dumbacher, John P., et al. 1992. Homobatrachotoxin in the Genus *Pitohui:* chemical defense in birds? *Science* 2258: 799–801.

Flannery, Timothy F. 1998. *Throwim Way Leg.* Text Publishing Company, Melbourne, Australia.

Frith, Clifford B., and Bruce M. Beehler. 1998. *The Birds of Paradise.* Oxford University Press, Oxford.

Gilliard, E. Thomas. 1969. *The Birds of Paradise and Bower Birds.* Natural History Press, Garden City, N.Y.

Gressitt, J. Linsley. 1982. *Biogeography and Ecology of New Guinea.* 2 volumes. W. Junk, The Hague.

Lovejoy, Thomas E., and Lee Hannah (eds). 2005. *Climate Change and Biodiversity.* Yale University Press, New Haven.

Marshall, Andrew J., and Bruce M. Beehler (eds.). 2007. *The Ecology of Papua.* Periplus Editions, Singapore.

Mayr, Ernst. 1941. *List of New Guinea Birds.* American Museum of Natural History, New York.

Mittermeier, Russell A., Patricio Robles Gil, and Cristina G. Mittermeier, 1997. *Megadiversity—Earth's Biologically Wealthiest Nations.* CEMEX, Mexico City.

Mittermeier, Russell A., et al. 1999. *Hotspots.* CEMEX, Mexico City.

Mittermeier, Russell A., et al. 2002. *Wilderness.* CEMEX, Mexico City.

Mittermeier, Russell A., et al. 2006. *Lemurs of Madagascar.* Conservation International, Washington, D.C.

Nettle, Daniel, and Suzanne Romaine. 2000. *Vanishing Voices.* Oxford University Press, New York.

O'Hanlon, Redmond. 1997. *No Mercy: a Journey into the Heart of the Congo.* Alfred A. Knopf, New York.

Rand, Austin, L. 1936. The distribution and habits of Madagascar birds. *Bulletin of the American Museum of Natural History* 72: 145–495.

Rasmussen, Pamela C., and John C. Anderton. 2005. *Birds of South Asia: The Ripley Guide.* Smithsonian Edition, Washington, D.C.

Ripley, Dillon. 1942. *Trail of the Money Bird.* Harper Brothers, New York.

Ripley, S. Dillon. 1952. *Search for the Spiny Babbler.* Houghton Mifflin, Boston.

Ripley, S. Dillon, Bruce M. Beehler, and K. S. R. Krishna Raju. Birds of the Visakhapatnam Ghats, Andhra Pradesh. *Journal of the Bombay Natural History Society* 84: 540–559; 85: 90–107.

Senanayake, F. Ranil, and Bruce M. Beehler. 2000. Forest gardens—sustaining rural communities around the world through holistic agro-forestry. *Sustainable Development International 2000:* 95–98.

Terborgh, John W. 1992. *Diversity and the Tropical Rainforest.* W. H. Freeman, New York.

Wallace, Alfred Russel. 1869. *The Malay Archipelago.* [1962 reprint], Dover Publications, New York.

Whitmore, T. C. 1975. *Tropical Rain Forests of the Far East.* Clarendon Press, Oxford.

Index